PRAISE FOR D.

'Fascinating... A kind of cross
Records and the most engaging
Ka-Boom! is an absorbing journ
and smallest, the fastest and the slowest, and the blackest and the brightest.'

Daily Mail on *Ka-Boom!*

'Darling provides entertaining answers to questions you never even knew you had. You will learn something new in every chapter, on every page and probably in every paragraph. Hugely entertaining.'

Kit Yates on *Ka-Boom!*

'Delightfully described with an understated humour... This book is a fact-nerd's heaven... so much more than a list of interesting facts and figures. Read up on the fascinating extremes of the universe and entertain your guests at dinner parties!'

BBC Sky at Night on *Ka-Boom!*

'A glorious trip through some of the wilder regions of the mathematical landscape, explaining why they are important and useful, but mostly revelling in the sheer joy of the unexpected. Highly recommended!'

Ian Stewart on *Weird Maths*

'A wonderful new book... if you love journeying into imagined mathematical worlds and simply exploring, then [this book] is pure, unadulterated escapism... brilliant.'

New Scientist on *The Biggest Number in the World*

'Gripping... dazzling tales of madness and derring-do.'

Brian Clegg on *Mayday!*

'A hugely enjoyable roller-coaster aerial ride in the company of oddballs and heroes, complete with vertiginous frights, dances of death, lonely impulses of delight and acts of mindless masochism in the name of science.'

Jonathan Glancey on *Mayday!*

ALSO BY DAVID DARLING

Ka-Boom!

The Biggest Number in the World

Weird Maths trilogy

Megacatastrophes!
Nine Strange Ways the World Could End

Equations of Eternity

and others

A PERFECT HARMONY

Music, Mathematics and Science

DAVID DARLING

A Oneworld Book

First published by Oneworld Publications Ltd in 2025

A CIP record for this title is available from the British Library

ISBN 978-0-86154-985-6
eISBN 978-0-86154-986-3

Typeset by Tetragon, London
Printed and bound in Great Britain by Clays Ltd, Elcograf S.p.A.

The authorised representative in the EEA is eucomply OÜ,
Pärnu mnt 139b–14, 11317 Tallinn, Estonia
(email: hello@eucompliancepartner.com / phone: +33757690241)

Oneworld Publications Ltd
10 Bloomsbury Street
London WC1B 3SR
England

Contents

A glossary of musical and scientific terms can be found on page 235.

Preface

MUSIC AND MATHS are endlessly entwined in symbiotic relationship, nourishing one another. Through music, an interplay of numbers is infused with passion and life. Girded by maths, music gains structure and sense. Both are universal languages, seemingly very different but, in fact, deeply connected. Science forms the triumvirate. When music is played it becomes part of the physical world: its sounds in thrall to the laws of acoustics, its ultimate purpose realised in the living tissue of our senses and brains.

The trinity of music, maths and science has been with us since before the dawn of civilisation – evident in bone and mammoth ivory flutes made at least forty thousand years ago. The pitches produced by an even earlier Neanderthal instrument, fashioned from the thigh bone of a cave bear, match four notes of the scale we use most commonly today.

A Perfect Harmony takes the reader on a musical odyssey through time, from the Palaeolithic to the present, and a deep dive into the links between notes and number, musical perception and physics. At heart, it explores how something that, ultimately, can be reduced to mere equations and wave patterns conspires to have such a powerful emotional impact upon us. All of this is told against a backdrop of the human forces that have shaped the development of music – religion, society and artistic expression – in different parts of the world.

We venture, too, beyond the confines of our own music to consider what forms it might take elsewhere in the universe or if developed autonomously (as it doubtless will be) by advanced artificial intelligence. In this way we'll come full circle to confront the very nature of music in terms of its mathematical and scientific underpinnings, and the effect it evokes in the mind of the listener.

CHAPTER 1

Prelude

BEFORE THERE WAS music there was – what? Silence? Vibrations in air, water and rock existed long before our species took its first step along the way to Vivaldi and Van Halen. Sound is older than the hills. Music may be a relative newcomer. But its origins extend much farther back than perhaps we normally suppose.

It isn't much to look at: part of the femur of a cave bear cub, about eleven centimetres long – roughly the width of your hand. On one side are four holes in a line: two of them complete, the other two partial where the bone has been broken at either end. A fifth hole lies on the other side. The bone came to light in 1995 during an archaeological dig at Divje Babe, a cave in northern Slovenia. It was unearthed next to a hearth used by Neanderthals fifty to sixty thousand years ago and may be the world's oldest surviving musical instrument.[1]

Sceptics have suggested that the holes in the 'Neanderthal flute' were made by a carnivore, such as a hyena or brown bear. But the bite of an animal would surely have splintered the bone, and the alignment of the holes is hard to explain except as the result of purposeful manufacture. The spacings of the holes, too, bear the hallmark of handcrafting because they're just right to let recognisable tunes be played. Replicas of the flute have been made on which, remarkably, a scale can be sounded that's indistinguishable from one we commonly use today.[2]

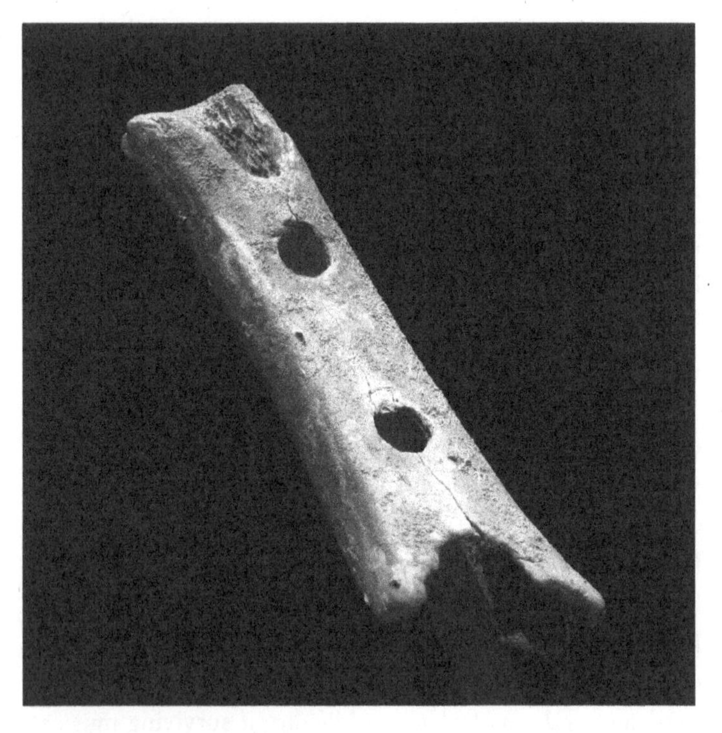

The Divje Babe flute, also referred to as the 'Neanderthal flute', measuring 11.4 centimetres in length. National Museum of Slovenia.

It's impossible to tell when music began. All we can say for certain, based on artefacts, is that by a few tens of thousands of years ago, humans and possibly their close relatives were fashioning instruments on which musical notes could be played. The controversial Neanderthal find aside, several bone and ivory artefacts have turned up in caves in Germany that are beyond doubt simple end-blown flutes dating back more than 35,000 years. It isn't hard to imagine a group of Stone Age men, women and children huddled round a fire as darkness falls while a haunting melody reverberates from the walls of their rocky shelter.

Musical instruments were among the first artistic tools to be manufactured. But long before they appeared, there was the human voice. Any sounds we can make today, the earliest members of our species were anatomically equipped to make equally well. A prehistoric baby transported to the present day from that far-off time eventually could learn to speak and sing as well as you and I.

By around 300,000 BCE – the date of the oldest known modern-human remains – our ancestors had the means to vocalise in a musical way. We can't be sure whether they actually did so, but there's every reason to suspect they may have explored such tones. Mothers make gentle sounds to soothe their babies. If those sounds are varied in pitch and calming to the ear, they amount to a simple, improvised lullaby. Very basic singing like this may have appeared on the scene long before any sophisticated spoken language. After all, it's much easier to produce a tuneful murmuring of sound, such as humming or crooning to an infant, than to form the vowels and consonants of complex speech. But in both cases – song and language – two essential ingredients have to be in place: a physical apparatus that can produce a wide variety of sounds and a brain capable of exerting fine control over that vocal machinery.

None of our nearest living relatives – apes and monkeys – can talk or sing as we can. Yet we have much in common with them in terms of anatomy and intelligence. So, what's prevented them from coming up with at least a simple spoken language or rudimentary forms of music?

First, it's not true that our primate cousins lack any musical talent or, at least, latent musical abilities. The indri, a type of lemur, is well known for its distinctive calls, made by groups of animals which climb high into the treetops so they can be heard far away. Their 'songs' sound like the sliding vocal exercises known as sirens that singers use when warming up. They may not be music as such – more a form of long-distance signalling – but they reveal an ability to sustain and vary musical notes in a controlled way. Indris also have a sense of rhythm. The sounds they make are typically either the same length or half as long as the gaps between them. Male and females sing at different tempos, but with the same rhythm, and may end with a steady slowing down, or *ritardando*. Mated couples will often duet, seemingly to stake out their territory and reinforce its boundaries against encroachment.[3]

Lar gibbons, native to the forests of South-East Asia, are another breed of simian songster. Their melodic cries became the subject of a study at Kyoto University's Primate Research Institute. Scientists there wanted to find out how well the apes could control the tonal quality of their calls.[4]

When we speak or sing, air passes over our vocal cords causing them to vibrate at a certain frequency, determined by the length of the cords and how tightly they're stretched by the muscles attached to them. Whole-number multiples of this fundamental frequency give rise to different harmonics.

The resonant frequency – the natural frequency of vibration – of the vocal tract then determines which of these harmonics are emphasised. By altering the position of the mouth, tongue, lips and teeth, humans can rapidly adjust the resonant frequency of their vocal tract to make the many different sounds used in speech and song. Operatic sopranos are especially adept at this so-called resonance tuning. They learn to tune the resonant frequency of their vocal tract to the pitch frequency produced by the vocal cords. In this way they can amplify their voice at certain pitches, enabling it to cut through and be heard even above the sound of a full orchestra.

In the case of lar gibbons, two different ideas had been put forward to explain how they produced their calls. The first was that the gibbon's vocal tract simply resonates in tandem with its vocal cords such that the resulting sound is predetermined like that of a wind instrument. The second was that apes have a more human-like arrangement in which the vocal tract and cords can be resonated independently.

To test these ideas, the Kyoto researchers put a captive female gibbon in a large box with a mixture of air and helium. If you've ever sucked in the contents of a helium balloon, you'll know what it's like to suddenly have the voice of a cartoon chipmunk. The gas doesn't change how the vocal cords vibrate, but it does alter the resonant frequencies of the vocal tract. By recording the gibbon's calls in a helium-rich atmosphere, the scientists were able to separate the different contributions from the vocal cords and vocal tract. What they found was surprising: the animal could modulate her voice just like a human soprano, adjusting to the effects of the new gas mixture by selectively amplifying lower-pitched sounds differently from those of higher pitch.

It seems that human vocal anatomy isn't as unique as many once believed. We aren't special in the way we make sounds, only in the extent to which we can control those sounds with our brains. Fine neural control over the vocal system is evidently a key factor that separates us from our fellow species.

Did language or music come first, or did they develop together? Traditionally, it's been considered that music, and singing in particular, is an evolutionary spin-off of language. But a more recent theory suggests the opposite, that language evolved as a subset of music. Notice how we talk to babies in a voice that's very different to normal. Babies respond best to sing-song speech that goes up and down with drawn-out vowel sounds. We adopt this approach naturally, without thinking, exaggerating pitch variations and emphasising certain syllables. We know intuitively that this makes it easier for a baby to discern different sounds and, over time, to recognise and distinguish between words. It seems that we all have a sort of ancestral memory of the time when language evolved from song. At any rate, it's clear that language and music are strongly connected, and the reason other animals don't have our level of linguistic and musical skills is the same.[5]

The question as to why monkeys and apes can't speak goes back to Charles Darwin's theory of evolution by natural selection of the mid-nineteenth century. Darwin himself thought that non-human primates couldn't talk because they lacked the brainpower. But over time, anthropologists latched onto the notion that it was the vocal tracts of our primate cousins that were holding them back. Now, it seems Darwin may have been at least partly right after all.

Research carried out at Princeton University has added to evidence that apes and monkeys have the basic vocal apparatus needed to produce speech similar to our own. Neuroscientists trained Emiliano, a long-tailed macaque at Princeton's primate lab, to sit in a chair while they recorded an X-ray video of him eating, yawning and uttering a variety of vocalisations. By analysing X-ray stills from the video, the team was able to put together a collection of ninety-nine different configurations of the monkey's vocal tract, from larynx to lips.* They then compared these configurations with what would be needed to produce distinctive vowel and consonant sounds. Finally, the researchers fed Emiliano's vocal tract shapes into a computer program that simulates vowel and consonant production given different anatomical settings. They chose an English phrase, 'Will you marry me?', which contains several different vowel sounds, and ran it through a simulation of the monkey's vocal tract. Sure enough, Emiliano's faked matrimonial inquiry, if not exactly human sounding, was clear enough to be intelligible.[6]

In terms of anatomy, it seems macaques aren't far from being speech-ready. The same is true of other monkeys and apes, and even some non-primate mammals, which have a similar vocal apparatus to Emiliano's. The reason they can't speak is not that they can't form a reasonable approximation of the necessary sounds, but rather that they lack the neural wiring to put those sounds together to make meaningful words. Equipped with a human brain they'd be perfectly capable of holding a conversation with us.

---

* The cartilage of the larynx and bones of the neck, jaws and teeth show up well under X-rays as dark shadows. Soft tissues of different densities appear as various shades of grey, so the overall X-ray image of the vocal tract is quite clear.

We'll look in vain for physical evidence of when our ancestors began to speak or sing, because neural circuitry doesn't fossilise. Our talent for sophisticated vocalisation presumably emerged gradually over the past six or seven million years, starting from the time our hominin forebears first diverged from the common ancestor we share with the chimpanzee, our closest living relative.

The long-tailed macaque has a vocal tract capable of intelligible speech.

Obvious advantages came from having better linguistic skills as we evolved away from apes and became more human.

An ability to communicate with greater clarity and precision is vital when you're working as a team and relying on your wits rather than on strength or speed. Bigger brains could support a wider range of spoken and sung vocalisations – the beginnings of language and music – which was important for our survival. Yet, at the same time as this neurological development, there would have been subtle changes in anatomy that helped refine the human voice, making it clearer and more flexible. One such adaptation has recently been identified by researchers at Kyoto. After looking closely at the larynxes of forty-three species of primates, they found a significant difference between the voice boxes of apes and monkeys compared with that of humans. Our larynx lacks a vocal membrane – a structure consisting of fine, ribbon-like extensions of the vocal cords. We also don't have balloon-like air sacs in the larynx that help some primates make raucous and resonant calls. The result of these laryngeal simplifications is that, although we don't have the vocal power to make ourselves heard across great distances, we're highly articulate and able to utter long, stable sounds with excellent pitch control.[7]

It's remarkable how far a few million years have taken us. But all of our linguistic and musical skills have developed from less-refined abilities already present in monkeys and apes. We often hum while preparing or eating a meal. So, too, it turns out, do gorillas. During an expedition to the Republic of Congo, scientists from the Max Planck Institute recorded two groups of western lowland gorillas in the wild and found that they hum and even sing during mealtimes. When gorillas hum it's like the 'mmm' that humans make, a steady low-frequency tone. Their food songs consist of a series of mismatched notes, like someone making up a random ditty on the spot. The inventiveness is what's interesting – the fact that the animals don't repeat the same song but seem to extemporise as they go along.[8]

In captivity, all gorillas occasionally hum or sing while eating – louder if it's a favourite food and each with their own distinctive voice. But in the wild it's a different matter. There, only the dominant silverback male sings, evidently to inform the others in his troop that it's time to tuck in.

Chimpanzees display another essential aspect of music: rhythm. They can drum and dance in time. We all know how a song with a strong beat will get our feet tapping and bodies swaying: there's an obvious close connection between the auditory and motor areas in our brain. Chimps, it seems, feel the groove too – clapping, swaying and bobbing their heads – when tunes are played for them. Evidently, the prerequisites for music are deeply rooted, extending back to a time long before humans appeared. Even birds, to whom we're far more distantly related, can produce tuneful connected sounds and, in the case of some parrots, will bob and weave to the beat of a pop song.[9]

Music is a form of communication. In the non-human animal kingdom, sounds and behaviours that we might loosely call musical – birds 'singing', gorillas humming and the like – serve some purpose related to survival. That purpose might be attracting a mate, warning off competitors or messaging within a group. Some of these more basic, animalistic urges are still evident in our music today – sexual signalling among them.

Led by Western Sydney University, research aimed at exploring the evolutionary origins of music found that boys singing in a choir alter their voices when girls are in the audience. Recordings were made of the St Thomas Boys Choir of Leipzig, with and without girls present for the recitals. These showed that the basses (postpubescent boys with deeper voices) sang with more energy in the formant range – the frequencies at which the vocal tract resonates – when performing with girls present. Both males and females, listening to the recordings later, could tell when the formant was enhanced, especially the high-energy peaks that usually coincided with vowels. But only females consistently showed a preference for the enhanced formant, regardless of the piece being sung. Human chorusing, it seems, serves not only to communicate socially but also as a subtle form of sexually motivated one-upmanship.[10]

Like some other animals, we employ music, in the broadest sense, to communicate cooperatively and competitively at the same time, at both group and individual levels. But we also use it for much more. Only with our own species, and perhaps with our immediate extinct relatives, has music been elevated to an art form. That happened when music began to be used to do more than just communicate but to convey emotions as well.

As soon as we start looking into the distant, prehistoric origins of music, we run into the more fundamental question: what *is* music? The answer seems obvious if you're listening to a Mozart piano concerto or even a child singing 'Happy Birthday'. We think we know what music is when we hear it, but an all-encompassing definition eludes us. Are the 'songs' of birds or whales musical? Birdsong certainly is, to the extent that it's been widely emulated in the works of composers – perhaps most famously in Beethoven's *Pastoral Symphony*, no. 6, op. 68. How about the crashing of waves on a shore or the tones of wind blowing through hollow logs? If not musical by intent or design, such sounds in nature surely served as inspiration for early human experiments in acoustic art.

At the heart of what most of us would think of as music are two elements: melody and rhythm. Melody can be any sequence of sounds that are controlled in pitch. Rhythm is a definite pulse or beat, which the sounds follow. Alone these two components are necessary but not sufficient; otherwise, a police siren would count as music. Another factor is subjective: how we're affected by what we hear. Music stirs an emotional response in us. A study by scientists at the University of California, Berkeley, mapped thirteen key emotions – among them joy, dreaminess, defiance, eroticism, beauty, triumph, fear and anxiety – that can be triggered by listening to different types of music.[11]

No two definitions of music are the same, because we're trying to force something that's partly a matter of opinion and cultural upbringing, and is fuzzy around the edges, into a box with a specific label. In some languages, such as Tiv, Yoruba and Igbo, spoken in parts of Nigeria, there isn't even a word for 'music'. But let's stay on familiar ground. According to Webster, music is 'the science or art of ordering tones or sounds in succession, in combination, and in temporal relationships to produce a composition having unity and continuity'.

That definition seems to rule out the possibility of music just 'popping up' in the natural world. No dawn chorus ever spontaneously organised itself into a recognisable tune, let alone a symphony or sonata. We talk about the songs of birds as if they were musical, but even among researchers who've analysed birdsong there's no agreement. Studies of animal 'song', whether of birds, mammals, frogs or insects, generally centre around how the sound is produced physiologically and its purpose. The closer a creature is to us in an evolutionary sense, the better are the chances it may give us some clues to the origins of music in our own species. But scientists are cautious about drawing any direct connections. To the animals themselves, their utterances are a means of signalling in one way or another, not an art form. Only humans, so far as we know, create and use music to entertain, convey feelings and influence mood.

Having said that, it's reasonable to assume that when people first started making music tens or hundreds of millennia ago, they found inspiration in the sounds around them, whether it was birdsong, a howling wind or a crackling fire. They would also have come across new ways of creating sounds as a result of toolmaking and random experimentation. These serendipitous discoveries doubtless played a part in the construction of primitive musical instruments.

The rhythmic striking of flint on flint when shaping stone axes or arrowheads, the beating of a stretched animal skin or the tapping of a hollow log or dried gourd would have given rise to early forms of percussion. Not that there's anything unique to humans about bashing things for an effect. An amazing variety of animals produce sonic vibrations, often by drumming parts of their body. The purpose might be to claim territory, boast superiority, signal readiness to mate or alert the rest of the group that a predator lurks nearby. And what better device than a loud, persistent rhythm to warn prey animals that they've been spotted and that an ambush is imminent?

Wild chimpanzees drum on the buttress roots of trees, sending out low-frequency booms that can travel a kilometre or more. Often they intersperse their percussive bursts with pant-hoots – the chimps' characteristic loud vocal calls. An international team of researchers, which followed and studied chimps in the Ugandan rainforest, found that the animals drum out messages to one another, each with their own signature style. The distinctive rhythms allow them to communicate over long distances, revealing where they are and what they're doing. Over time, the scientists themselves could often recognise who was drumming when they heard them. Each male chimp, they found, used a characteristic pattern of beats and drummed at different points in their call. Some individuals had a more regular rhythm, like a rock drummer, while others displayed more variability in tempo. One young male chimp, to whom the researchers referred as the 'John Bonham of the forest' – in homage to the great Led Zeppelin drummer – went in for extended solos with lots of rapid beats, clearly proud of his skills and keen to show them off.

The animals appeared to use their signature rhythms only when they were travelling, their long-distance signalling serving as a kind of personal beacon. The chimps' drumming in this way could solve a long-standing puzzle: why wild chimps greet each other when they meet but don't seem to make any gesture when they depart again into the forest. Given that they're effectively able to keep in touch while they're away, farewells aren't really necessary.[12]

The chimps' behaviour opens a window on how early humans would have exploited the possibilities of percussive sounds – first as a way of signalling, then as a form of artistic expression and, finally, if the drum beats were regular, to supply a rhythmic accompaniment for dance, chanting and other activities that promoted social cohesion.

As early humans developed simple tools, they would have discovered new ways of making musical sounds. Blowing across the opening of a hollow bone produces a note, and the longer the bone, the lower the pitch. Add holes along the length that can be covered and you have a primitive flute. The twang of a hunting bow would have led to the discovery that the sound varies depending on the length and tautness of the bowstring (made from animal gut). Arrowheads have been found in South Africa dating back sixty or seventy thousand years, so that by this time at least our ancestors had probably discovered the possibilities of vibrating strings as a source of musical tones.

Long before the first settlements were established then, some form of music and a range of primitive instruments – blown, stringed and percussive – had appeared. But as people began to develop agriculture and live semi-permanently in larger groups, the possibilities for musical development grew and music as we know it began to take shape.

CHAPTER 2

Airs of an Ancient Age

AMONG THE FIRST places where civilisation took root was the fertile plain between the rivers Tigris and Euphrates. Here, in Mesopotamia – what today is part of Iraq – some of the earliest depictions of musicians and their instruments have been found. Artefacts dating back to around 2800 BCE show lutes and lyres being plucked at ceremonies and banquets. Mesopotamians sang and made music with wind, string and percussion instruments – twenty-six different types are listed on one clay tablet from the twenty-sixth century BCE. Instructions for playing them have also been discovered on clay tablets and, in some cases, the instruments themselves have been found, including a silver flute and the oldest known string instruments – the so-called Lyres of Ur.[1]

Many preconceptions about the development of music have been blown away by archaeological research in the Middle East. Musicologists no longer accept the notion of a linear development of music – the idea that the further you look back in time, the simpler the music becomes and the fewer the notes upon which it was played. It was once taken for granted that the Mesopotamians must have had a relatively basic pentatonic, or five-note, scale. The assumption was that the Greeks inherited this system and then went on to invent the heptatonic, or seven-note, scale that dominates Western and much other music today. But in the 1960s came a remarkable discovery that overturned this dogma.

In the ruins of the royal palace of Ugarit, an ancient port city in what's now northern Syria, were unearthed fragments of a number of clay tablets inscribed in cuneiform. Several languages in the ancient Middle East used cuneiform – the earliest known type of writing, based on wedge-shaped symbols that were pressed into soft clay with a reed stylus. The tablets found at Ugarit, dating back to about 1400 BCE, were written in the language of the Hurrian people and contain parts of songs. One of them, the 'Hymn to Nikkal', is the oldest surviving, more-or-less complete work of notated music ever found.

The Hurrian songs reveal some extraordinary things about Mesopotamian music. Because they're specific compositions, they tell us a great deal about the musical system in use at the time. It certainly wasn't primitive or even pentatonic. Incredibly, not only was it heptatonic but also diatonic – involving five whole tones interspersed, more or less evenly, with two semitones per octave. Sing the familiar 'do re mi fa so la ti do' and that's a diatonic scale. So, the kind of melodies being played in Mesopotamia, at least three and a half thousand years ago, would have sounded – in terms of their typical sequence of pitches – much like Western European folksongs.[2]

Music was important in Mesopotamia long before the Hurrian songs were written down. Symbolic representations for 'harp' and 'musician' appear in the earliest known examples of writing – cuneiform-inscribed tablets from the city of Uruk dating to 3200 BCE. The first harps were almost certainly derived from the hunting bow by adding a resonating chamber, or sound box, and strings that could be tuned. Together with lyres, they were the dominant musical instruments by the start of the third millennium. Lyres were developed from the harp by replacing the single bow shape with two upright arms joined by a crossbar. Their strings, instead of being connected directly to a sound box, were made to run over a bridge.

The Mesopotamians also incorporated a wide variety of drums and flutes into their music, but the lyre and harp overwhelmingly guided their approach to music theory. Fortunately, the remains of several of these instruments have been found at the royal cemetery of Ur, a prominent city-state in the south of the region. Being 4,500 years old, the wood in them had largely disintegrated, but the materials, including gold, silver, copper, lapis lazuli and mother of pearl, with which they'd been richly decorated or overlaid, had survived. The ceilings of the tombs had collapsed, smashing the instruments and mixing their various fragments together. But over many years, archaeologists and museum researchers have disentangled the debris. Painstakingly, they've matched up the decorations and reassembled the shattered pieces. As the forms of the original instruments emerged, replicas could be made and played to bring back sounds that hadn't been heard for over four millennia.

The most common string instrument of the mid-third millennium BCE was the bovine lyre, with a sound box shaped like a reclining or standing bull. The Mesopotamian Sun god Uta (later known as Shamash) was often taken to assume the form of a bull, particularly in his role at sunrise. From the graves at Ur, several examples of this design were recovered, ranging in size from handheld specimens to the magnificent

Bull-Headed Lyre, found in 'The King's Grave', near the bodies of more than sixty soldiers and attendants. The head of the bull is gold-plated over a wooden core with a lapis lazuli beard. A panel below the head depicts four early Mesopotamian funerary rituals. The first shows a man wrestling two bulls with human heads; the second, a hyena serving meat and a lion carrying a jar; the third, the lyre itself being supported by a bear and played by a horse-like animal (suggesting that two people were needed to operate the instrument in real life); and the fourth, a scorpion-man guarding the underworld. The head and panels are mounted on the front of the sound box from which rise two long arms and a crosspiece to hold the strings.[3]

The Bull-Headed Lyre of Ur in the Middle East gallery of the University of Pennsylvania Museum of Archaeology and Anthropology.

Playable reconstructions of several of the Ur instruments have been made. Their registers and resonances vary depending on size and produce sounds ranging from those reminiscent of a cello or bass viol to a small guitar. But what was actually played on these instruments? What combinations of sounds, plucked on the strings, might resemble ancient Mesopotamian music?

Fortunately, we don't have to guess. Among the many cuneiform tablets that have been recovered from the lands of Mesopotamia are a small number that give details of the tuning and playing of lyres and their instrumental relatives. Because of these texts, we know that by at least 1800 BCE, and probably well before, there were standardised methods for tuning any of seven different but interrelated modes based on a diatonic system. When this fact first came to light, in the 1960s, it astounded musicologists. These seven modes were the very ones that had previously been assumed to originate with the Greeks, some 1,400 years later. In fact, as we've seen, one of the modes is equivalent to our own major scale on which much of modern music, including pop music, is based!

Encoded in clay, almost four thousand years old, are instructions for tightening and loosening the strings of a lyre in order to tune it to each of the seven principal modes. The main intervals, too, are identified in cuneiform – fifths, fourths, thirds and sixths – together with names for the scales, the individual strings and the parts of the instruments. Even the fingering techniques to be used in playing are described in detail.[1]

Early Mesopotamian representations of lyres show them having anywhere from three to twelve strings. Several of those found at Ur have eleven and one has thirteen – more than enough on which to carry out the prescribed tuning methods using intervals of fourths and fifths and to accommodate the octave. Perhaps surprisingly, scholars haven't identified an actual word for 'octave' in either of the main languages used in ancient Mesopotamia – Sumerian or Akkadian. But it's clear that the concept was well understood because of the numbers used to represent the octave jump of the first and second strings – one instead of eight and two instead of nine. In other words, it was known that after ascending seven notes of the scale, the next would be a return to the starting point but an octave higher.

Human ears and brains haven't changed since the dawn of civilisation or, for that matter, since *Homo sapiens* first appeared on the planet. So, when Mesopotamians heard two strings being played that were tuned an octave, a fifth or a fourth apart, they immediately would have recognised, as we do, the sounds as being consonant. Just because people didn't have our technology or scientific know-how several thousand years ago doesn't mean they couldn't discern a pleasant musical interval or tune an instrument to play those consonances within a single octave. From the time in prehistory when some creative individual hit upon the idea of twanging two bowstrings at the same time, experimentation in tunings, intervals, consonances and dissonances would have begun, leading to certain universal musical truths.

Although we needn't be mystified to learn that the folk of ancient Mesopotamia used the same musical intervals and scales as we do in the West today, we still don't know the details of how instrumentalists and singers performed back then. The fact that they used multistring lyres and their ilk suggests that they sometimes would have played more than one string at a time. Indeed, the intervals and tuning instructions described in the cuneiform texts always refer to string pairs, or dichords, suggesting that plucking two or more strings simultaneously was common.

Knowing the tonality of the Mesopotamian scales, as we do, it's not hard to recreate the overall kind of sequences of notes that the instruments of the time would have been used to produce. Based on diatonic modes, in which the distances between one note and the next are pretty evenly balanced, the melodies wouldn't seem strange alongside many tunes you could hear on the radio. What we don't know is the style adopted by performers. Did they, for instance, stay on pitch or did they vary the pitch intentionally as happens in modern Middle Eastern music? This much is clear: the people of this most ancient civilisation knew of and used the same musical scales that we do now in the West. The music we associate with the Middle East today, with its exotic (at least to Western ears) scales, modes and microtones, developed, as we'll see, independently and in quite a different place.

Music played a pivotal role in Mesopotamian religion, and the origin of that deep connection between music and religion isn't hard to find. The people of this region were agriculturalists and animists – everything to them was alive and suffused with a supernatural essence. It's a belief they inherited from their Neolithic ancestors who saw music as a link with the voices of spirit-animals. In fact, many surviving pieces of Mesopotamian art show animals playing musical instruments and priests wearing animal costumes.

Through music, the people believed they could have a direct and intimate relationship with their many gods, some of whom controlled the great forces of the world. Rammanu, the thunder god, could destroy crops at a whim during storms. Enki, god of deep water, could flood the lands. Along with a host of other deities, they had to be worshipped and appeased. Music was an important way to do this. Each god or goddess had certain material associations. More than that, things and substances could become actual manifestations of deities if they were divinised. A temple statue, for instance, wasn't merely a representation of a god but a form in which the god lived and must be propitiated. This was also true of musical instruments. Rammanu's breath was present in the sound of a reed pipe. Enki, later known as Ea, could take the form of a kettledrum, the sound of which reflected his power and high status. Ninigizibara, a minor goddess, was associated with the *balaĝ*, a type of lyre.

An object, such as a statue or musical instrument, was divinised by performing an elaborate and lengthy ritual, after which a particular god would take up residence in the object. The process was directed and overseen by lamentation priests whose job was to keep the god in a good mood throughout. Details of some of these rituals have been preserved in texts and pictures. In one, a steer – a young, castrated bull – was chosen by divination and led to the temple on the appointed day. Offerings were made to Enki while incense was burned and incantations sung. Around the steer, twelve figurines of gods were placed in a special, magical arrangement. The animal's mouth was washed while incantations were sung into its ear through a tube of aromatic wood. Then followed the slaughter, accompanied by lamentation songs in which the bull was assured of immortality. The beast's heart was removed, sprinkled with juniper and burned. Its hide was treated with flour, wine, fat, alum and gall apples, and then applied, in a complex series of stages, to a previously prepared drum frame. Offerings were made to various gods to ensure their goodwill. On the fifteenth day after the drum's completion, it was presented to the temple god. Now a divine temple drum, it could be played only by the priest to whom it was assigned. Through these procedures, the bull was given life after death: its 'heart' survived in the drum, which wore its skin and continued to beat in the rhythmical beating of the instrument.

Because of their importance, temples employed highly trained, professional musicians and singers to perform hymns and play instruments during religious events. Singers are sometimes shown in iconographic representations standing next to instrumentalists, holding their hands against their chests or stomachs to assist in diaphragmatic breathing. Every city had a temple with its most important citizen as the precentor – the head of worship – who knew the intricacies of communicating with the gods. The temples, known as ziggurats, were massive, stepped, pyramid-like structures whose design often incorporated astrological and cosmological elements, aligned with celestial events. The steps linked the ground with the sky, serving as both a symbolic and a literal stairway to heaven. These great structures thus afforded a connection between the earthly realm and the divine, emphasising the importance of the gods in daily life.

Beyond religion, music found many secular uses in Mesopotamian society. For entertainment, musicians and singers were often hired to perform at banquets, feasts and social gatherings, like event bands today. Music was also a means of cultural expression and storytelling. Poems like the *Epic of Gilgamesh* were sometimes accompanied by musical performances that dramatised the tales of heroes, bringing legends to life. Emotionally elevated in this way, cultural and historical narratives flowed easily from one generation to the next. Each region and city-state had its own musical tradition and style, adding to the richness of Mesopotamian culture. Some texts mention the use of music to soothe and heal individuals suffering from physical or mental ailments. At the other extreme, music often loudly accompanied processions and marches, the beat of drums and the sound of trumpets boosting the morale of soldiers and creating a sense of unity and purpose during military campaigns. Every one of these ways in which music played an important part in people's daily lives would continue throughout the centuries, in other places and societies, right through to the present day.[4]

Music also permeated the education of Mesopotamians. Musical training was considered essential to a well-rounded upbringing, especially for students in the elite classes and those being prepared for temple duties. The preparatory knowledge needed grew still further from about 1000 BCE with the emergence of large ensembles and orchestras.

Trade and travel helped spread the Mesopotamian musical system, along with other cultural elements, to neighbouring lands and eventually as far as the Mediterranean coast. Along the way, there were developments by the Babylonians, Assyrians and Chaldeans. Even before 3000 BCE the trading connections between Egypt and the Middle East were strong, and Mesopotamian influences can be found in Egyptian art, architecture, technology, pottery, weaponry and religious imagery. We know next to nothing about ancient Egyptian music theory because no written records of it have been found, but it seems highly likely that it was based on diatonic scales and modes imported from the Mesopotamians.

Tradition held that the source of Western music, like much else of our culture, was ancient Greece. In both music and mathematics, Pythagoras and his followers appeared to have a central role. But Greece was part of a nexus of civilisations that included Egypt, Phoenicia, Mesopotamia and others. Almost everything in Greek art, music and philosophy has Eastern antecedents. When Pythagoras travelled to Egypt, as it seems he did in the sixth century BCE, he brought back many influences – among them elements of music theory most likely Mesopotamian in origin.

Pythagoras is one of the first mathematicians we hear about in school because of the famous triangle theorem that bears his name. He's also the first character usually mentioned in any chronology of music. Considering how often his name crops up in different contexts, it's easy to suppose we know a lot about him. In fact, we don't know when he was born or died, where he was educated or, for that matter, any specific details about his life. None of his writings have survived and much of what's been written about him is anecdotal and probably fictitious.

While Pythagoras remains a shadowy figure, there's no such mystery surrounding the school of philosophy he founded. Central to the teachings of Pythagoreanism was the concept of number. Whole numbers were revered by the Pythagoreans and relationships between them held in high regard. Followers of Pythagoras, if not the man himself, grasped the importance of ratios between small whole numbers in music.

Pluck a string that's been stretched tight and can vibrate freely between two points on a hollow wooden board. Call this the root note or tonic. Now, hold down the string at its midpoint and pluck it again. The new sound is exactly an octave higher than the first.

Sing 'do re me fa so'. At 'so' you've reached the fifth note of the scale, which is to say, you've jumped up by an interval of a fifth. Played together, the root note and the fifth sound pleasant and free from any tension. Find the point along the stretched string where this consonant interval – a fifth – sounds and you'll notice that the ratio of the length of the vibrating part of the string to the non-vibrating is 3:2.

As we've seen, string instruments have been around a long time. The Lyres of Ur predate Pythagoras by two thousand years – almost as wide as the gap that separates us from the ancient Greeks. A lithograph on a wall of the Trois Frères cave in France, dated to about 13,000 BCE, shows a figure dressed as a bison who appears to be using a hunting bow for making music. The players of these ancient instruments would have discovered early on, by trial and error, the various relative lengths of string that sounded intervals that went well together. But the Greeks were the first, as far as we know, to explore in depth the intimate connection between music and maths.

Pythagoras and his followers built an entire cult around the belief that 'all is number' and that whole numbers were especially significant. Each of the numbers one to ten had a special significance and meaning to them: one was the generator of all other numbers, two stood for opinion, three for harmony and so on, up to ten, which was the most important and known as *tetractys* because it's the triangular number made from the sum of the first four numbers, one, two, three and four.

In music, the Pythagoreans delighted in the fact that the most consonant intervals corresponded with the simplest whole-number ratios. A vibrating string held down at its halfway point (2:1) sounds an octave higher than when open. Held down and played so that the lengths of the vibrating to the non-vibrating sections are in the ratio 3:2 gives a perfect fifth, 4:3 a perfect fourth and 5:4 a major third. In modern terms we'd say that the frequency – the number of vibrations per second – varies as one over the string length. Halve the length of string and you double the frequency of sound it produces. So, the ratios just mentioned, for the octave, fifth, etc., also apply to frequency. For example, in jumping by an interval of a fifth, the frequency increases in the ratio 3 to 2, by a fourth in the ratio 4 to 3 and so on.

The Pythagoreans were captivated and enthralled by this realisation that simple ratios of vibrating strings corresponded to harmonious intervals. It tied in so well with their fundamental conviction that the universe was based on whole numbers. And so they were led to a grand vision that in the heavens was a perfect marriage of music and mathematics. Borne on transparent celestial spheres and moving in circular paths were ten objects, in order from the centre: a counter-Earth, Earth itself, the Moon, the Sun, the five known planets or 'wandering stars' (Mercury, Venus, Mars, Jupiter and Saturn) and, finally, the fixed stars. The separations between the spheres corresponded to the harmonic lengths of strings, so the movement of the spheres gave rise to a sound (inaudible to human ears) known as the harmony of the spheres. This sound was supposedly carried by a fifth element, in addition to the terrestrial quartet of earth, air, fire and water, known as aether or quintessence.[5]

The Greeks, as we've seen, inherited their knowledge of the heptatonic, diatonic scale, and its various modes, from the East, ultimately from Mesopotamia. They weren't the first to come up with the system of seven notes, and their corresponding intervals, with which we're familiar today. But in other ways they were innovative musical theorists. The most consonant of the diatonic ratios (apart from the octave) – the perfect fifth – is the basis for what's become known as Pythagorean tuning. This was used by musicians in the West until about the end of the fifteenth century CE, when its limitations for playing a wider variety of pieces became apparent. We'll look at Pythagorean tuning, and why it was replaced, in Chapter 4.

The other major musical contributions of the Greeks were two new families of modes – the chromatic and the enharmonic – which we'll also explore in more detail in Chapter 4 when we look at how scales have evolved, and in Chapter 12, when we plunge into microtonality. For now, think of anything that's in a chromatic mode as sounding somewhere between modern Middle Eastern music, with its exotic twists and turns, and jazz or blues. Chromaticism happens when notes that don't belong in a diatonic scale are injected into a composition or performance. Some chromatic modes have more of a jazz or blues vibe, while others, like the so-called Lydian chromatic, are definitely Middle Eastern in flavour.

Enharmonic modes are even further removed from the kind of music we're used to hearing. These can't even be played on an instrument which produces only diatonic notes, such as the piano. Modern Western scales use only tones and semitones, but ancient Greek theory allowed the octave to be divided into smaller divisions, known today as microtones. These microtonal modes had their heyday in the Classical period of Greece (from about 510 to 323 BCE) and fell from favour in the subsequent Hellenistic period, which lasted until about

30 BCE. Like the chromatic modes, they largely disappeared from the Western music scene altogether but were revived in the Middle East by the Arabs, where they continue to thrive, as well as in other parts of the world.

Over the past fifty or sixty years, a major assumption of musicologists has been turned on its head. The music we tend to think of as being uniquely Middle Eastern in flavour, with its sinuous turns and twirls and unexpected notes, is a direct product of the West – of ancient Greek modality. A scale known as the double harmonic major is stereotypically taken to be Middle Eastern in origin: it's the one that Hollywood always chooses to accompany scenes of camels crossing the desert or tales of Arabian nights. But actually it's straight out of the Classical Greek playbook. The exotic sounds of the Middle East first emerged in Greece, two and a half thousand years ago. On the other hand, the wellspring of modern music in the West, based on diatonic modes, wasn't Greece at all but Mesopotamia – a region home to modern-day Iraq, Kuwait, Turkey and Syria.[6]

The ancient civilisations of the East and of Greece and Rome gave us a rich palette of scales and modes from which to create endlessly different melodies. But one aspect of music was almost entirely absent from these early developments. Until about a thousand years ago, notes were generally played one after another. There was rhythm and melody but little or no harmony. Then, in Europe, the first experiments began in this new direction – vertical rather than horizontal – in which notes were sounded together to extraordinary effect.

CHAPTER 3

Towards Harmony

SONGWRITERS, COMPOSING A new song today, will often start with a sequence of chords, maybe strumming them on a guitar or playing them on a piano, and then come up with a melody line that fits with that chordal progression. Such an idea would have seemed bizarre to anyone in the ancient world and, in fact, is alien to anyone who creates or habitually plays music that's non-Western.

Whenever two or more different pitches are sounded at the same time, we're in the realm of harmony. Think of music as having two dimensions: horizontal and vertical. Melody consists of single notes played strictly one after another. Together with rhythm it can be thought of as extending horizontally in time. Melody and rhythm can exist perfectly well without harmony and, indeed, much of the world's music throughout history has been essentially non-harmonic.

Harmony forms the vertical dimension – the simultaneous sounding of different pitches – and is what most obviously differentiates Western music from other traditions over the past millennium. The West has almost an obsession with harmony, to the extent that it dominates every aspect of music theory. There's also a tendency to think that this heavy usage of verticality means that Western music is somehow more complex and advanced than older, melodic traditions. Nothing could be further from the truth. Many highly sophisticated musical styles, such as those of India and China, consist basically of unharmonised melodic lines and their rhythmic organisation. Listen to an Indian *raga* or a singer performing in melismatic style a microtonal tune from the Middle East, and you'll be left in no doubt that purely melodic music can be just as technically intricate and challenging as a Western piece with rich harmonies. But if we're looking at the evolution of music in the West – Western Europe, in particular – it's the development of the vertical dimension of sound that's most striking.

Throughout the Middle Ages in Europe, music, like everything else, was powerfully influenced by the Catholic Church. Monks and nuns composed and performed sacred music to create ethereal sounds that spoke to the faithful. The Church appropriated much of its cosmology from the Greeks but with a Christian deity infused into the notion of the harmony of the spheres. Music became an important way of paying homage to God and was expected to follow certain strict rules that would enable a listener to be receptive to spiritual thoughts. Melodies were kept pure and unaccompanied, lacking harmonies or chords as embellishments. Any form of perceived dissonance was frowned upon as a secular distraction – or worse, the work of the devil.

From about the third to the ninth century CE, Christian church music consisted solely of plainsong – a large body of liturgical chants, sung in Latin and based on eight different diatonic modes, similar to those used in ancient Greece. The verses might be sung by a soloist, or alternately by a soloist and a choir, or by a choir and a congregation. But even when many voices were involved they'd all be in unison. The term 'Gregorian chant' is sometimes used, but this refers to only a certain type of plainsong and stems originally from the mistaken notion that Pope Gregory I, in the sixth century, was a key figure in its development.[1]

With no room allowed for improvisation, plainsong demanded that the order and relative pitch of notes be learned orally and accurately. But this was a tough job given the hundreds of songs in the repertoire that had to be performed at various times during the Church year. A little relief came with the invention of neumes – an early form of musical notation (the topic of Chapter 5). Singers still needed to learn melodies by listening to them repeatedly but now, at least, had a way to jog their memory as to whether the next note was higher or lower than the previous one.

Around 800, some modest experiments in harmony began to intrude into the austere world of plainsong. The result was *organum*, in which the main melody of a pre-existing chant, such as 'Ave maris stella' ('Hail, star of the sea'), was supplemented by a second voice singing the same melody but a fourth or a fifth higher. The practice may have started as a way of adding greater emphasis to a chant or of reinforcing the sound to carry through the larger churches that were being built at the time. Previously, plainsong had been exclusively monophonic, with every voice singing exactly the same note, or an octave apart, at any given time. But with the introduction of other consonant intervals sung in parallel with the principal melody came the first stirrings of polyphony – literally, 'many sounds'.[2]

Perhaps inevitably this small step beyond strict monophony led to further invention and diversity. Sometime in the tenth century, compositions began to appear in which there was a separate melodic line, moving note for note in the opposite, or contrary, direction of the main chant. This new style was called free organum to distinguish it from the original parallel organum, and it came with an interesting challenge. The way the counter-moving melodic lines interacted made it impossible to maintain the accepted consonances of fourth, fifth and octave all the time. In free organum, it was always arranged that pivotal moments, including the start and end of phrases and key words in the text, were still perfectly consonant. But in between were other intervals, such as thirds and sixths, which, by the standards of the time, were distinctly dissonant in flavour.

Dissonance creates a sense of tension – a state that can be relieved only by movement back to consonance. Free organum is an early example of this motion from repose to tension and back to repose, which is basic to Western harmony. The emphasis on consonances at the end of compositions set the final points of arrival in strong relief and reinforced the idea of the cadence, or the finality of the keynote of a mode on which pieces normally ended. Dissonance, by contrast, added interest and drama – a touch of the unexpected – to music that could otherwise sound a little bland and predictable.

Organum always had a strained relationship with the Church because it involved a shift away from a single pure-melody line. The tension and release that became aspects of free organum compounded the uneasiness the Church felt towards dissonance. To the powers that be in the Catholic hierarchy, dissonance implied dissidence – a revolt against official policy and, by implication, the word of God.

But a revolution, at first slow and cautious, had begun. Although the Church remained highly influential in music until as late as the eighteenth century, with free organum surfaced the first hint of harmonic complexity. Gradually, more voices were added singing multiple independent melodic lines. Spearheading this phase of creativity, in its early days, were composers such as Hildegard von Bingen, a German abbess and polymath, who wanted not only to express their spiritual beliefs but also to explore the mysteries of the universe through the vehicle of music.

In her *Divine Harmonies*, Hildegard described music as the surest way of understanding the cosmos and God's plans for entering paradise. The tenth child of a noble family, she was gifted to the Church as human tithe at an early age. Sent to an isolated hilltop monastery in the care of an older girl, Jutta, she spent nearly forty years with a handful of other women from privileged backgrounds, each shut away for much of the time inside a small stone cell in an area of the monastery well away from the monks. Jutta eventually became abbess and then took Hildegard under her wing, treating her like a daughter and teaching her Latin, singing and the duties of a choir nun. After Jutta died in 1136, Hildegard herself took on the role of abbess. A few years later she started to write about visions she'd had as a child and continued to have. Her notes were transcribed by a friend, a monk called Vomar, and eventually caught the attention of Church leaders. In 1147, Pope Eugenius III, having been presented with parts of her book titled *Scivias* (from *Sci vias Domini*, 'Know the ways of the Lord'), the first of three describing her visions, declared her prophetic writings to be authentic and important.

Given this papal stamp of approval, Hildegard began to attract hordes of followers and became a force to be reckoned with in the power circles of twelfth-century Europe. For a woman to have licence to speak and write openly about religious affairs was incredibly unusual at the time, and Hildegard wasn't slow to grasp the opportunity. In an overwhelmingly patriarchal society, she used accounts of her visions and mystical experiences, sanctioned at the highest level, as a way to challenge authority and traditions. In 1150, she founded her own monastery at Rupertsberg. When critics complained that she'd allowed several dozen nuns there to dress elegantly in white with their hair worn long, she replied that she was simply following God's instructions. In a vision, she claimed: 'I saw that a white veil to cover a virgin's hair was to be the proper emblem of virginity.'

Hildegard became both adviser to, and approved critic of, kings, queens, emperors and popes. She wrote extensively and, like a modern-day celebrity, went on speaking tours. She also broke with tradition in her musical compositions, which were extraordinarily progressive – almost dangerously so – for their time. The more than eighty songs she wrote for the nuns of Rupertsberg often featured wide vocal leaps, intricate ornamentation and soaring melodic lines. She made liberal use of the Phrygian mode in conjunction with a drone note, which imparts an otherworldly quality to her pieces. On a piano, start on an E and hit all the white notes up to the next E and you'll have played a Phrygian scale. The second note, which is only a semitone above the tonic, gives an exotic feel – so exotic that, in ancient Greece, pregnant women weren't allowed to hear tunes written in the Phrygian mode for fear that the dissonance between pitches might summon demons![3]

It's hard to imagine anyone except Hildegard getting away with publicly performing a song such as her *O virtus sapientae* ('Oh strength of wisdom') in an ecclesiastical setting. The tune is in E Phrygian with a drone on the E that clashes spicily every time an F, the second note of the scale, is sounded in the melody. Bishops and priests must have winced at such a dissonant juxtaposition, but what could they do? Hildegard had the ear of popes and the firm belief that her compositions were inspired by God. To her, musical sounds were more than a mere combination of voices and instruments. They represented the balance of body and soul – the interconnectivity of humanity and the universe.[4]

Hildegard died in 1179. Over in Paris around this time, the magnificent new cathedral of Notre Dame was soaring skyward. This was to become the focal point for a remarkable further flowering of music, the likes of which had never been seen before. Hildegard had been a one-off creative genius who, single-handedly, pushed the envelope of mediaeval monophony. Notre Dame cathedral became effectively a research centre and testing-ground for true polyphony – the music of multiply independent melodic lines.

The names of relatively few significant composers are known from the Middle Ages up to this point. Hildegard is one, Kassia, another female composer who predated Hildegard by three centuries, is another, and then there is Pérotin of the Notre Dame school. We know little about the man himself but plenty about his music, which has come down to us in detailed written form. Pérotin explored the possibilities of unbridled polyphony for the first time. What would happen, he asked, if three or even four voices simultaneously sang different melodies, which overlapped to form a rich, intricate harmony? Thanks to advances in musical notation (explored in Chapter 5), he was able to write down interweaving melodic lines in detail so that vocalists could learn their respective parts and how they fitted in with the whole ensemble. Pérotin's adventurous exploration of interacting melodies led to the most elaborate combinations of intervals and rhythms ever heard in the West up to that time.

In Pérotin's compositions, we see an early stage in the development of what's known as counterpoint. This refers to when a number of melodies, which blend together harmonically but may differ in rhythm and pitch, are played or sung at the same time. The simplest examples are repetitive rounds, such as 'Three Blind Mice' and 'Frère Jacques'. Among modern popular songs that can be combined contrapuntually are 'My Way' and David Bowie's 'Life on Mars'.

Needless to say, the kind of musical extravagance in which Pérotin and others were starting to indulge – combining multiple melodies – didn't always sit well with Church authorities. In 1322, Pope John XXII issued a decree from the papal palace in Avignon forbidding all but the simplest kind of harmonisation in sacred music. It specifically deplored singers who 'truncate the melodies with hockets [sharing a melody line between two or more voices], deprave them with discants, and even trope the upper parts with secular songs'.[5]

Although the great musical innovations of the Middle Ages were hatched in a sacred setting, there was no shortage of secular music. Minstrels, singing of love and distant lands and events, historical or imagined, were employed by wealthy families or royalty to perform at banquets. They might be accompanied by the likes of harps, flutes and flageolets, but the music typically consisted of a simple melody with a narrow range and perhaps a little improvisation. Other minstrels were the equivalent of today's gigging musicians or buskers, usually solo but sometimes in pairs or small groups, who wandered from castle to county fair playing for whatever fee they could attract.

As time went on, new breeds of secular musician appeared. In the twelfth century, the tradition of the troubadour was born in Andalusia, the southernmost region of Spain. Troubadours (and their female counterparts, trobairitz) came originally from noble families and, unlike minstrels, wrote and performed poetry set to music for its own sake, not because they had to for a living. Later troubadours came from other, less privileged backgrounds and would travel from one royal or noble court to another, seeking patronage. Some were lapsed priests or students, whose prior knowledge of music and history gave them plenty of material for their compositions.

The term *ars antiqua* is used to refer to European mediaeval music between about 1170 and 1310. This spans the period of the Notre Dame school of polyphony and subsequent years, including the development of the motet – the most advanced form of multivoice polyphony. It was followed by what's known as *ars nova* in fourteenth-century France, the lowlands of the Netherlands and other parts of Europe. In this 'new style', the emphasis was on greater freedom and variety of rhythm and melody, supported by further progress in notation, in contrast with the strictness of music that had gone before. No longer did composers feel constrained to build on existing chants but focused instead on original polyphonic writing, both secular and sacred.[6]

The towering figure of the *ars nova* period was French composer Guillaume de Machaut. Such was his dominance of the style that musicologists now often use his death, in 1377, to mark the transition to the final stage of mediaeval music, *ars subtilior* (the 'subtler art'). His six-movement *Messe de Nostre Dame* ('Mass of Our Lady'), featuring four voices, is regarded as one of the crowning glories of Western music before the fifteenth century. The bulk of his output, though, was secular. He wrote a lot of songs about courtly love, combining his brilliance as both poet and polyphonic composer. Before him, composers wrote either songs, such as those of the troubadours, or church music. Machaut did both. Not only that but, in his motets, he often combined sacred Latin texts in the tenor (lowest)* voice with secular French texts in the upper parts – the kind of thing Pope John XXII was referring to when he denounced those who 'trope the upper parts with secular songs'.

Among other leading figures of *ars nova* were the composer Philippe de Vitry and the mathematician, astronomer and musical theorist Jean des Murs. Both sought connections between musical harmony and the cosmos at large. The Pythagorean idea that musical harmonies were inherent in the spacing of heavenly bodies not only persisted but was also investigated in much greater depth in the later Middle Ages. The ancient philosophy of *musica universalis* found its way into the quadrivium, a quartet of subjects – arithmetic, geometry, music and astronomy – taught in the upper division of mediaeval European

---

* In mediaeval polyphony, the chant is always said to be in the 'tenor' voice even though it's the lowest part, which might be sung by an alto or bass singer.

universities. Students were shown what were believed to be the strands joining music and astronomy: music expressing the beauty of simple numerical proportions to the ears, astronomy to the eyes. Through different senses, they were held to express the same underlying unity based on mathematics.[7]

For most mediaeval scholars, who believed that God created the universe according to geometric and harmonic principles, science – particularly geometry and astronomy – was linked directly to the divine. To seek these principles, therefore, would be to seek God.

Educated at the Sorbonne and a master of the quadrivium, Jean des Murs wrote in the 1320s about the deep connection between mathematics and music, and the perceived relationship between planetary movements and consonant intervals. He also worked with a group of music theorists, several of whom shared his interest in astronomy, to introduce new methods for measuring time in polyphonic music. His contemporary, Philippe de Vitry, an influential composer of *ars nova*, was a close friend of Nicole Oresme, one of the great natural philosophers and mathematicians of the fourteenth century.

Oresme built upon the Greek discovery that musical consonances correspond with certain simple numerical proportions. He then developed the theory of 'harmonic numbers' to include all numbers whose proportions could form musical intervals. Much of this mathematical work fed into the development in *ars nova* of rhythmic modes – basically, skeletal patterns of how a melody is situated in time.

Oresme believed that harmonic numbers formed the basis for all music that was possible. However, he drew a clear distinction between music which could actually be heard on a human level (*la musique sensible*) and that which was inaccessible to the senses (*la musique speculative*). Speculative music was an intellectual experience of the order that existed in the structure of the universe – the music of the spheres. Although this music couldn't be sensed directly, it was assumed that actual music could open the mind to speculative music and prepare the soul for its contemplation. In a wider sense, this was the underlying purpose of the quadrivium: to allow the student a glimpse of the divine by revealing the interplay between number, geometry, the movement of celestial bodies and the unheard music of the cosmos.

In his *Livre du ciel et du monde* ('Book of Heaven and the World'), Oresme also looked at a range of evidence suggesting that Earth might not be at the centre of the Solar System as the Church adamantly insisted. He pointed out that if Earth, and not the celestial spheres, were moving, then all the motions seen in the heavens by astronomers would appear exactly the same as if the spheres were rotating around Earth. He also noted that it would be far more economical for Earth to spin on its axis than that the immense sphere of the stars should rotate around our small world. But he backed away from making the bold – and perilous – claim that these arguments were fatal to a geocentric scheme of the universe. 'Everyone maintains', he wrote, 'and I think myself, that the heavens do move and not the Earth.' It would be another 150 years before the tide of opinion shifted in favour of a Sun-centred scheme.[8]

Old habits die hard, especially when an intimidating force like the mediaeval Church is there to reinforce them. Until the late fourteenth century the attitude towards musical consonance, especially among continental composers, followed, with few exceptions, the Pythagorean ideal. Only fourths, fifths and octaves – the intervals expressible as the simplest of numerical ratios – were considered uncontroversial for use in both melody and harmony. But in England the interval of the third (from 'do' to 'me') had been in common use for some time. A type of early English polyphony known as gymel, in which the voices move parallel to each other at the interval of a third, existed in the late twelfth century. In the famous song *Sumer is icumen in* ('Summer Has Come In') of the thirteenth century, the earliest example of a round and a remarkably elaborate piece for its time, the harmonic style is almost entirely centred on thirds. The sixth (from 'do' to 'la') was also widespread in English music. These two intervals, the third and sixth, which seem perfectly respectable and pleasant to us today, had been considered mildly dissonant throughout the rest of Europe. It took time to appreciate and tolerate the fact that they sound sweeter than relentlessly using all-too-perfect fourths, fifths and octaves.

By the early fifteenth century, the third and sixth had become accepted in European music as consonant intervals. Visits to the courts of northern France by the illustrious English composer John Dunstable played a part in this. There was also an increasing awareness of tonality – the concept of developing a composition with a definite keynote used as a point of departure at the beginning and as a point of arrival at the final cadence.

This was a time, too, when composers were starting to think more and more of harmony as a vertical phenomenon – to regard the sound of notes heard simultaneously as a definite entity. Although the basic style of composition was still essentially linear, consisting of intertwining melodic threads, the chords that emerged from the coincidences of notes in contrapuntal lines took on a personality of their own. A phenomenon that bears this out is *fauxbourdon* (French for 'false bass'), known in England as faburden. This was a musical style in which three voices moved parallel to one another. The middle voice consisted of a succession of notes in parallel organum a fourth below the top voice, and the lowest voice paralleled the sequence a third below the middle voice, producing a chord such as G–B–E, known as a 6/3, or a first inversion, chord. Originally an English development, *fauxbourdon* was adopted in the fifteenth century by continental composers seeking to enrich their harmonies. It combined the continental fondness for 'pure' intervals, such as the fourth, with the English taste for parallel thirds and sixths.

Today, every beginner guitarist or pianist learns chords almost at the outset. The commonest of these are known as triads, because they consist of three notes – the root, the third and the fifth – played together. Many well-known pop songs consist of progressions of just three triads. 'Leavin' on a Jet Plane', for instance, consists of a tune sung over the G, C and D major triads. 'Get Back', by the Beatles, is built on a chord sequence of just A, D and G. Strange as it may seem, the concept of a triad in music theory was unknown in mediaeval times. But the fifteenth century did see the rise of so-called tertian harmony, which allowed intervals like thirds and sixths to be used more freely. That was a significant change and it's what makes the music of composers like Dunstaple and Guillaume Du Fay sound noticeably more modern than anything that had gone before.

It's important to recognise that, as the mediaeval age drew to a close, the new vertical structures involving thirds and sixths weren't viewed as triads or even as single chords. Their movement and resolution was still dictated primarily by horizontal melodic motion and the rules for combining individual voices. Composers were just obeying the existing rules of counterpoint, but when doing this with third-based sonorities, some patterns emerged that eventually led to modern tonal harmony. We'll continue this journey of harmony in Chapter 6 when we move into the Renaissance era. Before that, though, we have some unfinished business to attend to that will take us, once again, back to the heady days of ancient Greece.

CHAPTER 4

A Question of Scale

IT'S EASY TO pick out a simple tune, like 'Twinkle, Twinkle, Little Star', on a piano or a guitar. The notes are already laid out on the keyboard or fretboard and you just have to play them in the right order. But it's taken thousands of years to arrive at the scales, and the tuning of individual notes, on which most Western music today is based.

Our familiar seven-note, diatonic scale had its origins, as we saw earlier, at least as far back as ancient Mesopotamia. It then travelled through time and space to Greece, where the Pythagoreans got deeply involved with linking the received scale to mathematics and to the universe at large. The consonance of the perfect fifth, produced by stopping and playing a monochord – an instrument with a single string – two-thirds of the way along its length, held a special fascination for them. It became the basis for so-called Pythagorean tuning, which held sway in European music until the start of the sixteenth century.

Not everyone, though, even in ancient Greece, was convinced that the notes of a scale should be fixed in terms of a mathematical ratio. One of these critics was Aristoxenus of Tarentum (a region of what's now southern Italy), who lived in the fourth century BCE. Of the 453 books he wrote on subjects that included philosophy and ethics, only 2 have survived, in partial form, to the present day, and they're both on music. His *Elements of Harmonics* and *Elements of Rhythm* are the main sources we have on ancient Greek music. Unlike the Pythagoreans, who were less concerned with music itself than about its role in a mathematically coherent universe, and unlike Plato and Aristotle, who were mainly interested in music's effect on human character, Aristoxenus examined music – in particular, its pitch structure – as a system in itself. He's the first known authority on pure musical theory in the classical world. In *Elements of Harmonics*, he argued that what matters when constructing a musical scale is not the mathematical relationship of the notes, or how they fit in with some grand cosmological scheme, but how they sound to the ear.[1]

Aristoxenus may have been a master of his subject, but it was the Pythagoreans who won out in having their system of tuning dominate Western music for the next two thousand years. To see how it works, keep in mind that a Pythagorean scale is constructed entirely from 'pure' perfect fifths and octaves. We'll use the modern names for notes, A to G, and also focus on the frequencies of notes (measured in hertz, or cycles per second), about which the Greeks knew nothing, to help us understand what's going on.

Let's construct the scale of C chromatic in Pythagorean tuning across a single or 'basic' octave. A chromatic scale, remember, means going up a semitone at a time, including all the notes within an octave from the starting note – equivalent to all the black and white keys in that span on a piano.

We start at the root note or tonic, C, and we'll choose, specifically, middle C, which in modern tuning, known as equal temperament, has a frequency of 261.6 hertz (Hz). The first thing to do is go up by a fifth, which means multiplying the frequency by 3/2. This gets us to 392.4 Hz and the note G on the Pythagorean scale. To produce the next note we ascend another fifth by multiplying by 3/2 again. Starting from C, we've now multiplied by 3/2 × 3/2, or 9/4, which is bigger than 2 and therefore takes us into the next octave. We want our scale to be built within the span of a single octave. So, we divide by 2, because all the frequencies in the starting octave are half those in the octave above. From our G, we've now gone down by a fraction of 3/2 × 1/2, or 3/4, which is equivalent to an interval of a fourth. Overall, from the C root note, we've multiplied by 9/8 and arrived on the D of our scale. Next we go up another fifth by multiplying by 3/2 again – 27/16 in total from the starting point – which brings us to the note A. To create a new note in the scale we always move by a fifth from an existing note. But sometimes we drop *down* by a fifth, which means multiplying by 2/3. If this takes us into the octave below, then we *multiply* by 2 to get back into the basic octave. For example, from the C root note, multiplying first by 2/3 then by 2, or 4/3 overall, brings us to the note F – the perfect fourth in the key of C. In the maths of music, going down a fifth then up an octave, or up a fifth and down an octave, is the same as going up by a fourth or down by a fourth, respectively.

By repeatedly going up or down in fifths, and adjusting, if necessary, to get back into our single-octave range, we obtain all the notes in our C chromatic Pythagorean scale. If you listened to this scale and then to C chromatic played in equal temperament, you wouldn't notice much difference. The frequencies of the notes in both cases are very similar. But the Pythagorean scale *is* different in one important respect. Two of the notes – the G flat (G♭), or diminished fifth, and the F sharp (F♯), or augmented fourth – are enharmonic. This means they're the same note with different names. On a piano, or on any instrument tuned to equal temperament, a G flat and an F sharp (in the same octave) have exactly the same frequency. But in Pythagorean tuning they differ in frequency by about a quarter of a semitone. The difference is known as a Pythagorean comma. It comes about because, in Pythagorean tuning, the ratio of frequencies between successive notes of the chromatic scale aren't the same. So, going up six semitones from the root note and going down six semitones from the octave above the root note doesn't get you to exactly the same place.

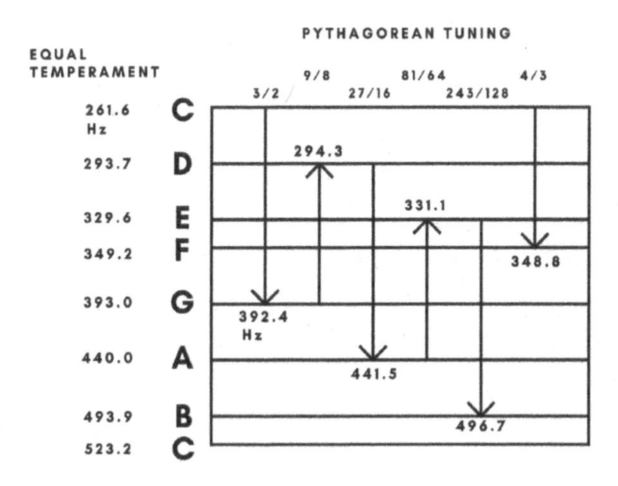

The notes of the Pythagorean scale of C major compared with C major in equal temperament. Each note of the Pythagorean scale is produced by jumps of a pure perfect fifth, corresponding to a frequency ratio of 3/2, from the previous note, with adjustments to bring all notes into the same octave. There are small but significant differences in frequency with the corresponding equal temperament scale.

The Pythagorean comma also crops up if we simply stack intervals of perfect fifths, one on top of another. After twelve hops of a fifth in equal temperament, we end up on our starting note but exactly seven octaves higher – roughly the entire width of a full-size piano keyboard. But the same number of hops by a perfect fifth in Pythagorean tuning takes us to a note seven octaves higher that's very noticeably dissonant with the note from which we began. In the key of C, we end on a B sharp that's about a quarter of a semitone below C. The mathematical reason for this is that no stack of 3:2 intervals (pure perfect fifths) will fit exactly into any stack of 2:1 intervals (octaves).

Although a quarter of a semitone may not seem much, it sticks out like a sore thumb in practice. If you were to play two notes at the same time, separated by that interval, it would sound terrible. The way around this problem is simply to leave out one of the competing discordant enharmonic notes – say, the diminished fifth (G flat in our case) – so that the resulting chromatic scale has only twelve notes with eleven fifths between them. Ten of those fifths will be in the exact ratio 3:2, but the remaining one will be so badly out of tune as to be unplayable. A horribly discordant interval such as this is called a 'wolf interval' (in this case a 'wolf fifth') because it produces a sound evocative of a howling wolf. It's one of the reasons that, beginning in mediaeval times, musicians began to experiment with other systems of tuning in an effort to avoid the dissonances that Pythagorean tuning causes, especially when music started to involve a wider range of sounds, harmonies and instruments.[2]

In Greek mythology, Harmonia was the goddess of peace and harmony – fittingly, since her parents were Aphrodite (goddess of love) and Ares (god of war). The Pythagorean idea that musical harmonies were inherent in the spacing of heavenly bodies persisted throughout the Middle Ages. As we've seen, the philosophy of *musica universalis* ('universal music') found its way into the quadrivium, a quartet of academic subjects that was taught after the *trivium* (grammar, logic and rhetoric) in mediaeval European universities and was based on Plato's curriculum for higher education. At the heart of the quadrivium was the study of number in various forms: pure number (arithmetic), number in abstract space (geometry), number in time (music) and number in both space and time (astronomy). Following the lead of Pythagoras, Plato saw an intimate connection between music and astronomy: music expressing the beauty of simple numerical proportions to the ears, and astronomy to the eyes. Through different senses, they expressed the same underlying unity based on mathematics.

More than two thousand years later, German astronomer Johannes Kepler took the notion of a musical cosmos a step further. Kepler believed in astrology and was devoutly religious, as were many other intellectuals of his time, but he was also a key figure in the scientific revolution of the Renaissance. He's best remembered for his three laws of planetary motion, built on the foundation of accurate observations of the planets by Danish nobleman Tycho Brahe. Early in his career, Kepler was fascinated by the notion that there might be a geometric basis to the spacing of the planets. To the Sun-centred model of the Solar System proposed earlier by Polish astronomer Nicolaus Copernicus, Kepler, in his 1597 *Mysterium Cosmographicum* ('The Cosmographic Mystery'), added the idea that the five Platonic solids – the only regular, convex polyhedra in three dimensions – held the key to the spacing of worlds. By inscribing and circumscribing with spheres these solids in a certain order – octahedron, icosahedron, dodecahedron, tetrahedron and cube – Kepler believed he could generate the orbs within which the six known planets (Mercury, Venus, Earth, Mars, Jupiter and Saturn) moved. God, it seemed, might not be a numerologist, as the Pythagoreans believed, but a geometer.

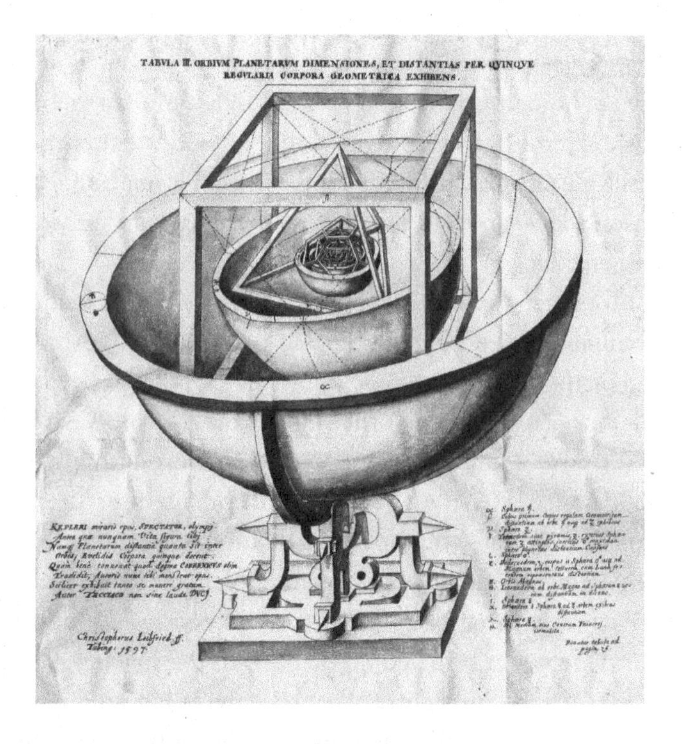

Kepler's representation of the heavens as a series of celestial spheres with Platonic solids filling the spaces between them.

Going beyond mere speculation, Kepler carried out acoustic experiments at a time – the dawn of the seventeenth century – when testing ideas in practice was still a novelty in academic circles. Using a monochord, he checked the sound made by the string when stopped at different lengths and established by ear what divisions were most pleasing. In addition to the fifth, which was of all-consuming importance to the Pythagoreans, he noted that the third, the fourth, the sixth and various other intervals were also consonant. He wondered if these harmonious ratios might be reflected in the heavens, so that the old notion of the harmony of the spheres might be brought up to date and more in line with the latest observations. Perhaps the ratio of the greatest and least distances between planets and the Sun matched some of the consonant intervals he'd found. But, no, they did not. He considered then the *speed* of the planets at the points of maximum and minimum distance, where he knew, from observations, that they moved the slowest and the fastest, respectively, in relation to the Sun. Movement, he noted, would be a better analogue than distance to the vibration of a string and, indeed, using this planetary property, he found what seemed to be a connection. In the case of Mars, the ratio of its extreme orbital speeds (measured in terms of angular motion across the sky) was about 2/3, equivalent to a perfect fifth, or 'diapente' as it was known until the late nineteenth century. The extreme motions of Jupiter differed by a ratio of about 5/6 (a minor third in music) and those of Saturn by very close to 4/5 (a major third). The corresponding ratios for Earth and Venus were 15/16 (roughly the difference between 'me' and 'fa') and 24/25, respectively.

Encouraged by these correspondences (which, it later turned out, are entirely coincidental), Kepler went in search of more subtle cosmic harmonies. He looked at the ratios of the speeds of neighbouring worlds and convinced himself that harmonious ratios underpinned not only the movement of planets individually but also how they moved relative to one another. He wrapped all of these thoughts on the subject into a grand unified theory of how consonant intervals in music were linked to movements in the heavens, and published it in his magnum opus, *Harmonices Mundi* ('The Harmony of the Worlds'), in 1619.

Shortly after, he made a discovery that today is known as the third law of planetary motion. He found a precise connection between the time it takes a planet to go once around the Sun and its distance from the Sun, namely, the square of a planet's period is proportional to the cube of its semi-major axis. This is the relationship still taught in physics classes today, but it was uncovered originally, and fortuitously, during the course of Kepler's mystical studies into the harmonic structure of the cosmos.

Kepler helped propel astronomy into the modern era with his crucial insight that the orbits of planets aren't circular, as had previously been believed, but elliptical. This paved the way for Newton's universal theory of gravitation but, less obviously, it set the stage for innovative and more flexible systems of tuning in music. From his experiments in auditory space, Kepler wondered if there was a smallest interval – a lowest common factor – from which all other harmonies could be built. He found that there wasn't. Just as planetary orbits weren't based on perfect circles, there was no neat and simple way to achieve musical consonance using one fundamental interval. This became most obvious when any attempt was made to change the key of a piece of music.[3]

Kepler lived from 1571 to 1630, a period during which music underwent an extraordinary transformation. Music was freed from its mediaeval constraints, polyphony evolved, harmony emerged based on chord progressions, compositions were created for larger ensembles and new instruments were developed. The inadequacies of Pythagorean tuning, with its overwhelming emphasis on the consonance of a single interval – the fifth – became ever more apparent.

Pythagorean tuning is one of a whole family of ways of tuning the notes of a scale known as just intonation. All just intonations have in common the goal of tuning musical intervals as whole-number ratios of frequencies. They differ only in how they go about achieving this. Pythagorean tuning is known as a three-limit just intonation because three is the highest prime number included in the ratios that define the intervals of its scale. The first few intervals of the Pythagorean major (seven-tone) scale, for instance, have frequency ratios of $9/8 = 3^2/2^3$ (major second), $81/64 = 3^4/2^6$ (major third) and $4/3 = 2^2/3^1$ (perfect fourth). There are also five-limit, seven-limit and other varieties of just intonation, some of which have been known since ancient times. 'Ptolemy's intense diatonic scale', for instance, discussed by Claudius Ptolemy in his book *Harmonikon* in about 150 CE, is a five-limit system that allows for a consonant major third, in addition to the perfect fourth and fifth of Pythagorean tuning. Notice how much simpler, in Ptolemy's scale, are the frequency ratios of the major third and sixth:

	Pythagorean tuning	Ptolemy's intense diatonic
Tonic	1	1
Major second	9/8	9/8
Major third	81/64	5/4
Perfect fourth	4/3	4/3
Perfect fifth	3/2	3/2
Major sixth	27/16	5/3
Major seventh	243/128	15/8
Octave	2	2

Pythagorean tuning worked well for mediaeval music, which utilised fifths and fourths almost exclusively. But over time the austere sacred music of the Middle Ages gave way to the sweeter, more varied sounds of Renaissance polyphony. As the use of major thirds and sixths increased, first in England in the 1300s, then in the rest of Europe, the need for these intervals to be more pleasing to the listener grew, pushing just intonations like Ptolemy's to the fore.[4]

Just intonation, of one form or another, dominated European music until around 1550. After that, its great strength – its focus on the beauty to the ear of using simple fractions to define intervals within a single key – increasingly became its weakness. Just intonation works fine providing we stay in the same key or use flexible instruments, such as the human voice, which can be easily retuned to adjust the pitch of notes. But any form of it runs into problems with instruments like the harpsichord or piano, which, once tuned, can produce only certain frequencies. As music entered the Baroque era and became more sophisticated, so did the need for scales that worked when a variety of instruments and changes of key were involved.

The first significant development beyond just intonation is known as meantone temperament. This emerged in Italy around the start of the sixteenth century during a burst of more sophisticated mathematical music theory. It's basically an adaptation of Pythagorean tuning with the goal of making the major third sound sweeter. An ideal, or 'just', major third has a ratio of 5/4, or 1.250, but the Pythagorean third is 81/64, equal to 1.266 – a bit too sharp. To tune the major thirds down, the interval of the fifth has to be shrunk as well. It turns out that the new fifths need to have a frequency ratio of $\sqrt[4]{5}$, or approximately 1.495. The fourth root of 5 is an irrational number, which means it can't be expressed in the form of one whole number divided by another. To the Pythagoreans, the use of such a monstrosity in their scheme for relating cosmic harmony to integers in music would have been anathema, totally unacceptable. But meantone temperament, by its inclusion of an irrational number, foreshadowed the next great development in tuning – the system that we still use almost universally today.

By Kepler's time, composers and musicians had started to break out of the rigid confines of just intonation. Even meantone temperament has limitations if playing outside a narrow range of related keys, or pieces that involve a lot of modifications to the pitch of notes, known as accidentals.[5]

A pioneer of an entirely new approach to tuning was about to emerge. He was the father of Galileo Galilei, the great Italian astronomer and physicist. Vincenzo Galilei advocated a twelve-tone scale based on what became known as equal temperament. In this system, every neighbouring pair of notes in a chromatic scale is separated by exactly the same interval, or ratio of frequencies. With twelve semitones the width of each interval is $\sqrt[12]{2}$ – the twelfth root of 2 – or approximately 1.059463. Multiplying $\sqrt[12]{2}$ by itself twelve times gives 2, which, in terms of frequency, takes us from the tonic to the same note but one octave higher. This arrangement means that none of the frequencies of twelve-tone equal temperament (12-TET) exactly match those of the corresponding notes in just intonation, except at the tonic and octave, although fourths and fifths are so close as to be almost indistinguishable. Equal temperament is a compromise: it isn't as pure sounding as just intonation but has the huge advantage of enabling music to be played that is acceptably harmonious in any key without the need for retuning. It made keyboard instruments, such as the piano, practical and musically flexible, and opened up broad new horizons in composition and orchestration.

One of the things you can do in equal temperament but not in Pythagorean tuning, or any other type of just intonation, is complete what's known as the circle of fifths. This is the musical equivalent of the periodic table in chemistry or a chart of elementary particles in physics. It connects a multitude of ideas and explains a great deal in a small amount of space.

To see how the circle of fifths comes about we'll choose as our starting note C, because the scale of C major is the only major scale that doesn't have any sharps or flats. A fifth up from C brings us to G (because if C is first, then D is second, E is third, F is fourth and G is fifth). Another way of thinking of this is that if C is the 'do', or root note of the scale, then G is the 'so'. If G is now taken as the root note, the fifth in the scale of G major is D. Another fifth up brings us to A, then to E and so on. When we get to F sharp we could continue up another fifth, which would bring us to C sharp. But the way the circle of fifths is normally drawn, we go back to C and then start to move the other way, in an anticlockwise direction, by going down a fourth. Going down a fourth from a note is the same as coming up a fifth from a note that's an octave lower, so the effect is the same.

By going down in fourths anticlockwise we eventually get halfway round the circle to meet up at the same point we reached in going up in fifths in the other direction. The note at the half-past position on the clock of musical notes can be called F sharp moving in one direction or G flat moving in the other. It's the enharmonic note we talked about earlier – the same note with two different names.

As you know, a circle is a closed shape. You can travel all the way round it and come back to your starting point. And the circle of fifths – or the circle of fourths, if you prefer – is a true circle with twelve points on it, representing the twelve distinct notes in the chromatic scale. The circle is closed because equal temperament guarantees that it's closed. The Pythagorean scale, based purely on fifths, fails to close the circle. In Pythagorean tuning, the circle of fifths becomes a spiral of fifths. And the effect of this spiralling is that you can never quite get back to the note you started on (or one that's several octaves higher) after completing a trip through all the other notes of the scale in ascending fifths. Equal temperament ensures that the frequency ratio between any note and the next higher note is always the same, and that twelve successive ratios returns us to a note that's exactly one octave higher than where we started.

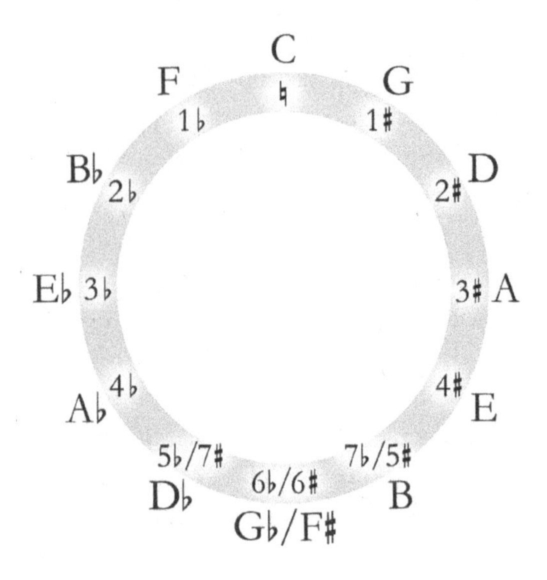

The circle of fifths.

Today, 12-TET is used almost universally in Western music but, like any tuning system, it represents a compromise. It offers tremendous flexibility in changing – or 'modulating' – between keys. For example, if you want to sing a particular Beatles' song, but some of the notes are a bit too high for your voice, you can easily transpose to a key that brings all the notes within your range. A disadvantage is that all of the intervals of a 12-TET scale are somewhat out of tune, some much more so than others. An equal-tempered perfect fifth is only about two cents – or two percent of a semitone – flat of pure, but an equal-tempered major third is about thirteen cents sharp of pure. For the most part, we don't notice this dissonance because we're so used to hearing music in the standard system. But compared with a pure third, which has a very stable sound, an equal-tempered third displays a perceptible beating. This instability comes about because of a mismatch between the tempered pitch and its untempered, embedded harmonics. It's a similar story for the other intervals of the 12-TET scale, although the major third stands out because of its musical importance.

In just intonations, some of the intervals are pure, which can result in a wonderful sound if the music is uncomplicated. But if the notes are pure relative to a particular starting note, they may be decidedly impure relative to other notes. As soon as music gets more complex, involving different key centres, chords or accidentals, any tuning system that involves different intervallic frequency ratios runs into problems.

Earlier tuning systems, with unequal intervals between notes, were less versatile but in many ways more interesting and colourful. Each key had a unique and distinctive sound to it. Prior to the twentieth century, those who made, performed or listened to music were aware of these differences and would associate different keys with specific emotional or qualitative characteristics. In his *Ideen zu einer Aesthetik der Tonkunst* ('Ideas Towards an Aesthetic of Music'), published in 1784, Christian Schubart offered a description of the characteristics of each key that proved influential in its day. D major, according to Schubart, is the key of 'triumph, of Hallejuahs, of war-cries, of victory-rejoicing'. F sharp minor, on the other hand, is 'a gloomy key: it tugs at passion as a dog biting a dress', whereas B major is 'composed from the most glaring colours – anger, rage, jealousy, fury, despair and every burden of the heart lies in its sphere'.

Subjective though these descriptions are, there's no doubt that each key does have a unique sound and character when the tuning system involves intervals of different sizes. Once equal temperament became the dominant tuning system, the aural quality of every key became the same, so these affective characteristics became mostly lost – but not entirely. It's true that equal temperament is now universally used for pop, rock, folk, jazz and for the vast majority of other music, including orchestral, heard widely around the world. But other tunings survive, especially in situations where the intent is to recreate the original sound of pieces from Baroque or earlier eras.

Traditional forms of music from other parts of the world, too, use different systems of tuning. This is partly why music of the East and Middle East has, to Western ears, an exotic quality. Arabic music, for instance, is based on 24-TET so that it makes liberal use of quarter tones. However, only a fraction of the twenty-four tones appear in any given performance, and these are determined by the maqam, or melody type, just as in Western music generally only seven out of the twelve tones appear, which are determined by the key. As in Indian *raga* and other traditional non-Western forms, there are strict rules, even within the most elaborate and protracted improvisations, which govern the choice of notes and their relationship, together with the patterns of these notes and the progression of the melody.

Our current system of chromatic equal temperament homogenised all twelve keys, removing the former system's individual key uniqueness and collapsing enharmonic pitches to a single fixed pitch. This has allowed music to become tremendously complex and sophisticated in certain ways, particularly with regard to harmony and movement through diverse keys.

From an early age, our brains become accustomed to the music that's pervasive around us, just as they adapt to the local language, the tastes of our home food and the ways of the people with whom we grow up. Music from other cultures may sound unusual and surprising, and yet for the most part, it is still pleasing to the ear. The different scales, intervals, rhythms and structures of musical pieces from other parts of the world may take some getting used to, but we almost always recognise them as being musical. This is because they, too, are based on acoustic patterns that can be reduced to relatively simple mathematical relationships that govern such elements as melody, harmony and tempo.

CHAPTER 5

Make a Note of This

IT'S SOMETIME IN the Dark Ages – the early mediaeval period – and you're a young monk or nun expected to sing in the choir of your little cloistered commune. Dozens of sacred hymns need to be memorised, all of them in plainchant. Learning the lyrics isn't so bad – they're just the words of the psalms and other sacred passages, every one of which is written down and you're supposed to know by heart anyway. But the melodies are another matter. There's nothing on paper (or parchment) to show you how the tunes go, so your only option is to listen to others singing the chants, over and over again, and then practise them till they stick in your mind – a task that could take years.

The importance of notation to the development of music is hard to overstate. But how do you represent – in written form – notes and their pitches, rhythms and all of the other details that describe a specific composition? We're so used to hearing recordings of music today and seeing sheet music or scores that reveal exactly what's supposed to be played and when, that it's easy to forget that it was all very different long ago: in the distant past, after a tune had been sung or played, nothing of it remained except what was held in people's memories.

The origin of spoken language, like that of music, is a matter for conjecture. Speech vanishes into thin air the instant it's uttered. Only when written languages appeared, after the dawn of civilisation, was there the potential for words and their meaning to be preserved for us to be able to decipher later on. The first city-states arose in Mesopotamia and the earliest written records stretch back to about that time, some five and a half thousand years ago. With writing came the possibility of capturing the lyrics of songs and setting down instructions for musicians.[1]

Who knows when the first attempts were made at saving the essence of a simple melody? Almost certainly, they're lost forever. The oldest known surviving written music is a collection of about three dozen songs, written in the now-extinct Hurrian language of the Middle East. They're inscribed in cuneiform on clay tablets found in the remains of the Royal Palace of Ugarit (present-day Ras Shamra), on a headland in northern Syria, and date to about 1400 BCE. Most are fragmentary, though one is reasonably complete: the 'Hymn to Nikkal', the goddess of orchards.[2] It consists of lyrics and instructions for both singer and nine-stringed lyre. Other tablets hint at how the notation system worked and how to tune a lyre. But mostly they offer only scattered clues, frustratingly imprecise. How amazing it would be to hear the original compositions played in that long-ago time.

The ancient Greeks had a form of musical notation, which was used for at least a thousand years, from the sixth century BCE, or even earlier, to about the fourth century CE. But surprisingly few examples of it have come down to us. The oldest are fragments of two compositions known as the Delphic Hymns, both dedicated to the god Apollo and apparently written for an important religious procession in Athens in 128 BCE. Both involve a single melodic line, but whereas the first uses vocal notation, in the second the notes are written with symbols for instrumental players.

We have to move forward in time, to about the first or second century CE, for the oldest entirely complete musical composition ever unearthed. The Seikilos Epitaph is a four-verse song, including Greek lyrics and notation, inscribed on a marble columnar tombstone found in the city of Aydin, near the west coast of what is modern-day Turkey. It's generally thought to be a dedication by Seikilos to his wife, Euterpe, though that's not certain because Euterpe was also the Muse of music. The piece takes under a minute to perform but is surprisingly complex and includes sections for both a harp and a flute-like instrument. The notation appears above the lyrics, which talk about the briefness of life and the importance 'to shine and not to grieve'. At the bottom was a message: 'I am a tombstone, an image. Seikilos placed me here as a long-lasting sign of deathless remembrance.' Unfortunately, these final lines were lost when, in a moment of Victorian madness, the stone was sawn off at the base to serve as a flowerpot stand for an early owner unaware of, or indifferent to, its significance.[3]

The Seikilos Epitaph is now more appropriately displayed in the National Museum of Denmark in Copenhagen. Various renditions of the song it bears have been recorded, differing somewhat according to how the notation is interpreted. It also pops up in the computer games *Civilization V* and Minecraft's *Greek Mythology Mash-up*, though greatly expanded in both from the original short melody into full-length musical tracks.

Fast-forward several centuries to early mediaeval Italy. Music, at this point, was still very much an oral tradition – songs had to be learned by ear. 'Unless sounds are held by the memory of man', said the scholar and archbishop Isidore of Seville, 'they perish, because they cannot be written down.' He wasn't alone in realising that this was a major stumbling block to musical progress. When Charlemagne became head of the Holy Roman Empire in 800 he set about trying to standardise chant. That meant coming up with a reliable way to transmit music across long distances. In short, an effective method was needed for writing it down. By the middle of the ninth century some progress had been made in developing a mnemonic system for recalling pieces of plainsong. Symbols known as neumes were described in *Musica Disciplina*, an important treatise by the Frankish music theorist Aurelian of Réome in about 850.

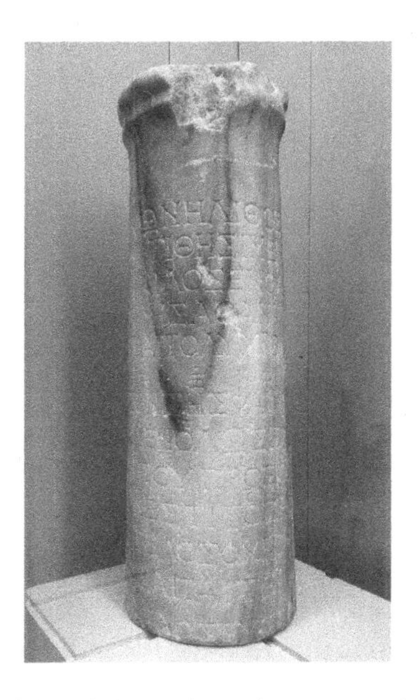

The marble Seikilos stele with poetry and musical notation. The National Museum of Denmark.

At first, neumes were no more than squiggles written above the text of vocal chants to indicate the general contour of the melody. The idea was to show if the notes in the tune went up or down over any given syllable. You can see how this would have been helpful to someone who already knew a song and just needed reminding how it went. But it couldn't teach a new tune from scratch, because it wasn't equipped to show exactly *how* high or low successive notes were supposed to be.

A thirteenth-century folio of the *Cantigas de Santa Maria* ('Canticles of Holy Mary') containing lyrics and neumes. The *Cantigas* consist of 420 poems with musical notation, written in the mediaeval Galician–Portuguese language during the reign of Alfonso X (1221–84). The music is written in notation similar to that used for chant but also contains some information about the length of the notes.

Then, in the tenth and early eleventh centuries, 'heightened' neumes began to appear, first in the churches of Benevento in southern Italy. These were a significant upgrade – often taking the form of little black squares placed at different distances above the words and syllables to show more clearly the relative pitches of notes. Now, for the first time, singers could see the whole melody in graphic form, with notes and the intervals between them plotted at consistent heights.[4]

The use of neumes spread to monasteries across Europe, and it wasn't long before further refinements were made to the notation. The next step was to add a line. This gave a reference point against which the notes could be compared – usually representing either the middle or the end of the scale. But with only one line as a fixed pitch level, there was still lots of room for ambiguity.

Early in the eleventh century, an Italian Benedictine monk and music theorist, Guido of Arezzo, stepped into the ring. Guido saw that people were struggling to learn chants from existing neumes and set about devising a more accurate system of notation. He gave the neumes a standardised, easy-to-read form so that each note had its own symbol. Then he drew four lines – a staff – onto which the notes could be placed and organised by pitch. One of these lines was coloured to serve as a point of reference for a specific named pitch so that singers could relate all the other notes to it. He also added time signatures and invented solfège – the framework we know today as 'do, re, mi, fa, so, la, ti, do'.[5]

Neumes placed on the staff showed exact pitches, allowing a singer to read an unfamiliar melody. News travelled slowly in those days and, even within Western Europe, somewhat different systems of neumes were used in different geographical regions. By about 1200, though, neumes had assumed the characteristic square shapes still used in the modern notation of plainchant. For the first time, musical ideas could be transmitted without the aid of sound, simply by marks on a piece of paper.

Still there was work to be done. Guido's staff was informative about pitch but said nothing about rhythm. The first real innovations on that front are often credited to the composers Léonin and Pérotin, of the Notre Dame school, who introduced what are called rhythmic modes. These were predefined patterns into which notes could fall. A note's length was determined by its location in a particular pattern.

It's no coincidence that these developments in notation came at a time when composers, such as Pérotin, were beginning to experiment with multiple lines of melody. Complex polyphonic pieces would have been impossible to arrange or learn without more precise methods for writing the pitch and length of notes in different voices. In Guillaume de Machaut's *La messe de Notre Dame*, which we first encountered in Chapter 3, four voices, each carrying an independent melody, come together to create specific harmonies at specific times. Marshalling such an extraordinary synthesis of sounds was only made possible through a paradigm shift: standard notation and precisely crafted scores became essential.

In earlier times, written music provided a mere guide, an aide-memoire. But from then on, written scores took on an increasingly important role. They never caught on for much of the rest of the world, where melodic, or horizontal, complexity often arose from improvisation and spontaneity. In the West, what ushered in written notation were the demands of vertical complexity. Improvements in notation enabled more intricate polyphony, which in turn encouraged ways to further refine the symbolic representation of the music.

Around 1250, Franco of Cologne, a German theorist, took music notation a step further. In his *Ars cantus mensurabilis* ('The Art of the Measurable Song') he proposed a system of symbols for different note durations. Previously, it was left to the singer to figure out the rhythms of a piece based on context. A stream of similar-appearing notes on the page would be interpreted as a series of long and short values by a trained singer based on a convoluted series of learned rules. Franco's symbols were imprecise by modern standards, but they nevertheless greatly expanded the options for conveying rhythm. They were further developed during the *ars nova* period by the addition of a wider range of defined note values and a push towards more clarity in notation. Composers were then able to deploy these more powerful tools for representing music with more intricate arrangements of pitches and rhythms.[6]

In 1320, French composer and poet Philippe de Vitry broke new ground, creating a system of mensural time signatures for minims, crotchets and semiquavers. By 1450, white notes had begun to overtake black notation, so most note values were written with white note heads – as today we'd write a semibreve or minim. Then, in the seventeenth century, the symbols for note values started to look a bit rounder. Throughout the seventeenth century, music notation continued to evolve to accommodate the music of Renaissance and Baroque composers. In particular, when instrumental music overtook vocal music as the most popular genre, a change in notation was needed. Instrumental musicians were still using Guido of Arezzo's system of staves and notation (albeit in slightly modernised form) but found there was no longer enough information for their liking. This led to the introduction of such details as bar lines, dynamic markings and performance directions. A fifth line was added to the staff, note heads changed from squares to ovals, clefs and signatures appeared, and dotted notes, flags and beams were devised to show arbitrarily small rhythmic values.

From the *ars nova* period on, the ability to lay out multiple lines of melody on a kind of musical spreadsheet freed composers to assemble far more complicated musical structures. Music could now be taken on a course to greater and greater sophistication: dazzling new possibilities for harmony began to suggest themselves. The rise of modern tuning also led to the evolution of a way to represent sharps, flats and naturals, known collectively as accidentals.

An early-sixteenth-century manuscript in mensural notation.

A thousand years or so ago in Europe, music effectively used only seven notes. These are roughly the ones now referred to as natural notes: A, B, C, D, E, F and G. But over time, people started to realise that there was more than one useful pitch between the notes A and C. Depending on the musical context, sometimes a lower-pitched B sounded better, and at other times a higher-pitched B seemed right. Fortunately, in those days, there were also a few different ways of writing the letter B, so people started representing the lower version of B using a round B, called a rotundum (b), and the higher B using a B quadratum or a square B. As time went on, people discovered more and more of these in-between notes, so the symbols that originally referred only to which B to play were repurposed as general symbols for whether or not a note should be played low (that is, flat) or high (sharp). The flat symbol (♭) is still pretty recognisable as a 'b', albeit with a pointy base. The square B, on the other hand, evolved into two different symbols – one to represent a note in its natural position (♮) and the other to mean a note raised, or sharpened, from its natural position (♯). So as a bizarre quirk of musical history, the sharp, flat and natural symbols are all just versions of the letter B from mediaeval times!

Around 1440, the German goldsmith Johannes Gutenberg developed the movable-type printing press, based on ideas and technology that came originally from China and Korea. Gutenberg's press ushered in the age of mass-produced books – and mass-produced sheet music. At the very time when music notation was becoming more sophisticated, allowing for more complex and expressive compositions, along came this marvellous device for disseminating written music to a wider audience.

With the printing of music came the need to standardise the symbols used in notation. Composers still wrote their music by hand in the first place and might go through many drafts while developing their works. Handwritten manuscripts in museums testify to the evolution of musical ideas, with sections often crossed out and new ones scribbled in. The finished composition would be passed to a copyist to produce parts for first performances before later being typeset for printing and public distribution. Printing enabled a much wider and faster sharing of ideas, and musicians and other composers could learn about the music of others without having to attend concerts of their works. The increasing availability of printed music also allowed music to be studied and analysed by students.

The first musical superstars emerged. In Renaissance England, Elizabeth I granted Thomas Tallis and William Byrd, his pupil at the time, a monopoly to print and publish music, which resulted in their works becoming widely known. Elsewhere in Europe the development of printed music helped to give composers a degree of independence from wealthy patrons, since they could now earn an income from publishing their own creations.

Many older ideas in music notation, especially from the *ars nova* period onward, were retained and refined even as progress in music propelled innovative ways to symbolically code the new developments. The Baroque period, from about 1600 to 1750, saw, as its name suggests, an outpouring of ornamentation, which in music took the form of melodic embellishments – trills, mordents and appoggiaturas – each requiring its own special sign.

The Romantic period, stretching from the dawn of the nineteenth century to the first decade of the twentieth, brought an increased focus on emotional expression and individuality in music. Composers began to include more detailed instructions for dynamics, tempo and articulation in their scores, allowing for greater control over the interpretation and performance of their works. From basic indications of a simple song line going higher or lower, the complexity of musical notation had grown to the point where it could now specify in detail all the parts for a 100-strong symphony orchestra and chorus.

Over the past hundred years or so, music notation has continued to evolve as composers, now freed from virtually all constraints of tradition, have explored extraordinary new sonic possibilities. So-called graphic notation, in which art and music are combined in a sort of musical map, offers the performer a loose guide, rather than strict instructions, on how to play the music. Graphic scores emerged as a reaction against the rigid constraints of detailed sheet music, using visual symbols, shapes and lines to represent musical ideas. They allow composers to convey concepts like texture, density and spatial relationships between sounds, and provide performers with vastly more interpretive freedom. The stirrings of this revolution can be traced to the late nineteenth century in the scores of Gustav Mahler, and even earlier, with those of Beethoven, which are full of scribbles, footnotes and improvisation marks as the composers strained to break free of the rules of conventional notation.

In the 1920s and 1930s, the largely self-taught American composer and pianist Henry Cowell pioneered a wave of avant-garde music that both broadened the scope of musical invention and occasionally offended audiences. Aged seventeen, Cowell enrolled at the University of California,

Berkeley, and studied under Charles Seeger, the renowned musicologist and composer. Intrigued by Cowell's radical approach to music, Seeger encouraged him to write about his methods and the theory behind his tone clusters – chords comprising several adjacent notes in a scale, for example C, C sharp and D, played simultaneously. Cowell later fleshed out his ideas and thoughts on transforming musical notation to accommodate his unconventional sonic explorations in a book, *New Musical Resources*, published in 1930.[7]

Still a teenager, Cowell wrote the piano piece 'Dynamic Motion' (1916), his first significant work to explore the possibilities of the tone cluster. It calls for the performer to use both forearms to crash massively dissonant chromatic chords and for the keys to be held down to extend the mingling of overtones from the cluster as the sound dies away.

During his first European tour in 1923, Cowell played at the famous Gewanhaus concert hall in Leipzig and received a memorably hostile reception from an audience unprepared for his musical adventurism. As he progressed further into the concert, deliberately saving the loudest and most provocative pieces for last, the crowd's mood turned ugly and more audibly aggressive. Gasps and screams were heard, and Cowell recalled a man near the front threatening to physically remove him from the stage if he didn't desist. 'Do you take us for idiots in Germany?' the outraged patron exploded.

In later life, Cowell became markedly more conservative in his musical direction following several years of confinement in the notorious San Quentin State Prison on charges related to homosexuality. But he'd already had an indelible effect on the emergent avant-garde music scene – no more so than through his pioneering graphic notation. Those he influenced included fellow composer Morton Feldman, who produced the first full-blown graphic score for his *Projection 1* (1950) for solo cello. Not only does the piece sound ahead of its time but the score looks more like a circuit diagram than anything resembling traditional written music.

Throughout the 1950s and 1960s, a new generation of postwar composers – Krzysztof Penderecki, Karlheinz Stockhausen, John Cage and Roman Haubenstock-Ramati, among them – began using graphic notation as a serious alternative to dots on staves. An extraordinary example is *Treatise*, written between 1963 and 1967 by the British composer Cornelius Cardew, who, it comes as no surprise to learn, trained as a graphic designer. Comprising 193 pages of lines, symbols and various geometric and abstract shapes, it gives no specific instructions on how to play the piece or what instruments should be used, but instead aims to inspire creativity in the performer and offer multiple choices of expression. Among contemporary musicians, Brian Eno has made extensive use of graphic notation. Having no formal musical education and being unable to notate in an orthodox way, he's made graphic scores a normal part of his process. His stock-in-trade being sound textures, these would, in any case, be difficult to notate in the traditional way.[8]

The twentieth century also saw the rise of aleatoric music (*alea* was a Roman dice game), which brought elements of chance and unpredictability into compositions. An early example of this approach came in the Baroque period, when composers would often use 'figured bass' notation. In their scores, they'd supply a bassline accompanied by numbers and symbols. These figures represented the intervals and chord qualities that should be played above the bassline to create harmonies. Musicians skilled in figured bass interpretation, known as continuo players, used these figures as a guide to improvise and fill in the harmonies according to the conventions of the time. Figured bass notation allowed for some degree of flexibility and creativity, making each performance unique.

Beginning in the eighteenth century, musical dice games caught on in Western Europe. The idea was to use dice to randomly generate music from precomposed options. One or two dice would be rolled and the corresponding number on a chart would be used to identify a particular measure of supplied sheet music. Some well-known composers of the time became involved. They included C. P. E. Bach with his 'Method for making six bars of double counterpoint at the octave without knowing the rules' (1758) and Maximilian Stadler's 'A table for composing minuets and trios to infinity, by playing with two dice' (1780). The most famous version of the game was published in 1793 and posthumously attributed to Mozart, though it's never been proven whether or not he'd actually been involved in devising it. *Musikalisches Würfelspiel* ('musical dice game') consisted of 272 musical measures and a table of rules for selecting specific measures given a certain dice roll. The result was a randomly constructed sixteen-bar minuet and sixteen-bar trio. Despite the fantastic number of possible combinations, the musical fragments were cleverly written such that any selection would produce a satisfying piece matching all the harmonic and compositional requirements of minuets and trios of that time.

In the twentieth century, some composers, such as John Cage, went even further, intentionally leaving compositions unfinished before using random-number generators or the roll of dice to fill in the missing parts. In his *Music of Changes* (1951), Cage turned to the *I Ching*, an ancient Chinese book of divination, to generate random numbers for representing tempo, dynamics and note duration. This served as inspiration for some classical composers to inject aleatory decision-making, to a greater or lesser degree, into their music. Pierre Boulez composed pieces that offered players a limited number of possibilities from which to choose – all of which were tightly scripted. *Concerto for 2 Pianos* by Alan Hovhaness (1954) combines elements of European fugue and Indian *raga* – a style that inherently involves improvisation. The score to Stockhausen's *Klavierstück XI* (1956) is printed on a single page and contains nineteen musical fragments to be played in a predefined order, although the performer may choose to start at any point in the cycle. *Pithoprakta* ('actions through probability') by Iannis Xenakis (1956) is conceptually inspired by the physics of gas molecules. Compositions like it, which incorporate physical or mathematical principles, are sometimes called stochastic music.

In indeterminately notated music, composers eschew the traditional five-line staff with treble or other forms of clef. Instead, they use graphic notation or text notation to give players guidance on what to do.

A small number of films have used aleatoric music as part of their musical score. John Williams's partially aleatoric score for Robert Altman's *Images* earned him an Academy Award nomination. *X-Files* composer Mark Snow occasionally took an aleatory approach in scoring the long-running TV show. Radiohead guitarist and avant-garde composer Jonny Greenwood has used aleatoric techniques in scores for Paul Thomas Anderson films.

With the advent of electronic and computer music, new notation systems were developed to accommodate the unique capabilities of these technologies. Composers have created notation methods for specifying parameters such as frequency, amplitude and modulation, as well as for programming sequencers and other electronic devices. The rise of digital sheet music and notation software also transformed the way musicians interact with scores, enabling dynamic, interactive and customisable representations of music.

We'll revisit the subject of notation in Chapter 12, when we delve into the multidimensional world of atonal music. But for now it's time to retrace our steps and go back several hundred years to see how scientific, mathematical and musical ideas evolved together after the Middle Ages to propel us towards an extraordinary future.

CHAPTER 6

Renaissance and Beyond

THE RENAISSANCE (literally, 'rebirth') – the period between about 1450 and 1650 – witnessed a spectacular flowering of art, science, literature and music in Europe following the relative stagnation of the Middle Ages. It was a time when, as one of the exemplars of 'Renaissance Man', Leon Battista Alberti, put it, 'A man can do all things if he will.' Absent was the specialisation we see today, as subjects flowed seamlessly into, and nurtured, one another. This great blooming of polymathic creativity and intellectual revolution had been a long time coming. But its seed had been planted nearly a thousand years earlier.

Between about 500 CE and his death in 524, the Roman senator, philosopher and all-round scholar Anicius Manlius Severinus Boethius produced a body of literature that had a profound effect on learning throughout the Middle Ages and into the early Renaissance. Boethius translated many classic works from Greek into Latin and also wrote a number of important treatises on music, astronomy, geometry and arithmetic – the classic quadrivium.

In his *De institutione musica* ('Fundamentals of Music'), Boethius focused heavily on pitches and musical intervals, but not simply from an academic point of view. Following the lead of Pythagoras, he saw knowledge of mathematical proportions in music as a way to attain virtue. He wrote:

> [W]ithout doubt music is naturally united with us that we cannot be free from it even if we so desired. For this reason the power of the intellect ought to be summoned so that this art, innate through nature, might also be mastered, comprehended through knowledge.[1]

The proportions of musical scales, Boethius insisted, revealed the order that underlay the human soul as well as the physical universe – the very rationale behind Creation. In 1500, a thousand years after it was written, Boethius's *Fundamentals* was still the essential starting point for any student wanting to study music as part of the quadrivium. But it was no longer the only work of musical theory in circulation. The fifteenth century had seen a huge expansion in the amount of music literature available in the West. This was thanks mainly to the efforts of humanists working mostly in Italy who'd scoured European monastic libraries in search of ancient musical texts. Many of these works came from the Byzantine Empire – the continuation of the Roman Empire in the Eastern Mediterranean until the fall of Constantinople (now Istanbul) to the Turks in 1453. They included new Latin translations of writings by Ptolemy, Euclid, Aristotle and Plato.

The Pythagorean tradition, as advocated by Boethius, still ran strong in the early Renaissance period, but theorists now had to deal with a much greater variety of ideas about the nature of music. Debates continued to rage around pitch and tuning, harmony and how these related to the human mind, the body and the physical universe. But added to these was the question of the *moral* effects that musical forms, particularly the various modes, produced on the listener.

Tales from ancient literature warned of, or celebrated, the effects that music could have on the individual. Renaissance theoreticians became fascinated with them. Frequently they quoted legends, like the one in which Pythagoras calmed a violent youth by changing a piper's tune, or Alexander the Great, who had been stirred to battle by a song written in the Phrygian mode.

In trying to come to grips with the wider, philosophical issues of music, Renaissance theorists also soon realised they didn't properly understand the pitches and tuning systems that had been used by the Greeks. Mediaeval music theory had naively assumed that the eight modes of plainsong – Dorian, Phrygian and so forth – were identical to the similarly named modes of the ancient world. It was only with the rise of humanism – a non-theistic worldview centred on science and reason – in fifteenth-century Italy that theorists gradually uncovered the truth about the various scales, modes and tuning systems that underpinned ancient music.[2]

The problem was that although Middle Age scholars took Boethius's *Fundamentals* to be effectively the bible of music theory, no one had really bothered to examine it closely. One of the first to do so was Frenchman Johannes Legrense, also known as Johannes Gallicus of Namur. As a youth in France, he was a 'cantor' (singer) but later became a *musicus* – a musician. In those days the two were considered totally distinct. Music research and teaching were far more prestigious occupations than composing or performing. The singer, or player of music, was thought of as the servant of the musician or music theorist. It was the musician's task to make value judgements in musical matters based on reason.

Legrense came to Italy and studied at a school for noblemen in Mantua. While there he was exposed to the teachings of Boethius and began to look deeply into the content of *Fundamentals*. As a result, he discovered a major – previously unsuspected – difference between the Greek systems of music and mediaeval plainsong. This isn't to say that Legrense was a heretic or revolutionary – far from it: while at Mantua he became a Carthusian monk and was a critic of both secular music and polyphony. But he opened the door to questioning the precise pitch relationships or intervals that had existed in ancient music.

Four years after Legrense died in 1473, Johannes Tinctoris, a Flemish music theorist, published his *Book on the Art of Counterpoint*. For the most part, this is an in-depth treatment of harmony and polyphony. But in the foreword, Tinctoris deals with the music of the spheres, putting forward arguments both for and against its reality. For much of the Renaissance, it was still popular to believe that celestial bodies made sounds as they moved and that there was a relationship between these sounds and harmonies on Earth. Some theorists even maintained that music should be studied for the changes it might bring about on Earth and among the stars. Tinctoris, though, came down on the side of Aristotle, insisting that the Sun, Moon and planets produced no sound as they travelled through space.

The issues raised by Legrense, Tinctoris and others in the early Renaissance began a battle for supremacy between two major schools of thought: scholasticism and humanism.

The former sought to reconcile classical teachings about the nature of the universe with Christian theology. It emerged in the monasteries of mediaeval times and formed the main style of education in the first universities. The scholastics' aim was to tie reason to faith – to mould the philosophies of ancient Greece to fit a universe in which the overarching truth was the existence of a supernatural God. Above all, scholasticism was a way of teaching, through analytical thinking and intellectual debate, but with the ultimate goal of validating a biblical view of the world.

Humanism, as the name suggests, is centred on the nature and importance of humankind. Humanists emphasised personal experience as the starting point for establishing knowledge. During the Renaissance, this didn't imply doing away with God: in fact, most humanists of this period were Christians. One of the earliest was the philosopher, mathematician and astronomer Nicolas of Cusa, who also happened to be a German cardinal. Although he was never accused of being a heretic, he wrote about Earth being in motion and not at the centre of the universe, and speculated that life might exist on other worlds. As a mathematician, he delved into unsolved problems like the classical one of squaring the circle.[3]

Much of the musical theory that developed in the first part of the Renaissance, until about 1550, was highly mathematical and technical in nature. There was an intense focus on rediscovering the precise pitch relationships that had existed in Greek music, even though it had little impact on music in practice. Most of those involved had a loftier goal than just pinning down the numerical ratios that underlay ancient tuning systems. They saw the mathematical study of pitch as a way of glimpsing the universal harmonies they believed inspired music's beauty, stirred the senses and perfected the soul. It was the musician's equivalent of the alchemist's quest for the philosopher's stone. Both sought to reach a higher consciousness – a sudden transcendent vision of the world *behind* the world.

Throughout the Renaissance and beyond, prominent mathematicians repeatedly delved into music theory. Marin Mersenne, often called the father of acoustics and well known for his work on prime numbers, wrote several treatises on music, including *Compendium Musicale* (1618), *Harmonicorum Libri* (1635) and *Traité de l'Harmonie Universelle* (1636). He also corresponded on music theory with other important mathematicians, notably René Descartes, who founded analytic geometry. In England, clergyman John Wallis, who helped develop calculus and invented the symbol for infinity, wrote on tuning and published editions of classical works on music by Ptolemy and the mediaeval scholar Bryennius. Later, Leonhard Euler, preeminent mathematician of the eighteenth century, expressed his musical ideas in a 1731 book whose title also serves as its own endorsement: *Tentamen Novae Theoriae Musicae Excertissimis Harmoniae Principiis Dilucide Expositae* ('An Attempt at a New Theory of Music, Exposed in All Clearness According to the Most Well-Founded Principles of Harmony').

As humanist-trained scholars began to challenge previously accepted notions about ancient music, the hot topic of the day was precisely what harmonic intervals and pitches had governed music almost two thousand years earlier? Many Renaissance theorists believed the answers might give clues to nothing less than the relationships that underpinned all

Creation and that governed music's influence on the human body, mind and spirit. Much of the theorising was abstruse and, ultimately, as misconceived as alchemical efforts to turn lead into gold. But out of the musical ferment some remarkable progress was made – developments that would influence the future sound and direction of Western music. At the centre of this transformation was one of the most distinguished theoreticians of the sixteenth century and his erstwhile student, the father of Galileo.

Born in 1517, Gioseffo Zarlino became a singer at the cathedral in his hometown of Chioggia and later its deacon and principal organist. Shortly after being ordained in 1540, he went to nearby Venice to study with Flemish composer Adrian Willaert, who at the time was choirmaster of St Mark's Cathedral. Zarlino eventually succeeded his teacher in that position and, in his book *The Art of Counterpoint* (1558), credited Willaert with reintroducing a sophisticated style of composition that he believed had been inspired by the ancients. Like those who'd come before him, Zarlino strongly believed that the correct tuning system to be adopted for polyphony and counterpoint should be based on small whole-number ratios. He advocated for this purpose Ptolemy's intense diatonic scale, which, as we saw in Chapter 4, brought the interval of a third into consonance and was therefore better suited to post-mediaeval compositions, which used a lot of thirds and sixths.

Not only was Zarlino the most respected musical theorist of his day but he was also largely focused on choral polyphonic sacred music. His *Art of Counterpoint* survived as a practical manual on the techniques of counterpoint for composers until well into the seventeenth century. But it was flawed in one obvious way. Ptolemy's scale is workable for the human voice and for instruments such as the violin, which can quickly adjust for the shifts in note spacings that happen when there's a change in key in just intonation. But it creates major problems for any instrument that's fretted or, like a keyboard, relies on fixed tuning.[4]

During the 1560s one of Zarlino's students was Vincenzo Galilei. Being a lutist, Galilei saw the obvious snag with Zarlino's push to have Ptolemy's scale universally adopted: you'd need a different lute depending on which key or mode a piece was written in. Zarlino argued that just intonation could serve the needs of polyphony because, after all, it was good enough for the Greeks. Galilei started corresponding about this with the historian Girolamo Mei, the first European after Boethius to have carried out a detailed study of ancient Greek music theory. Mei pointed out there was no evidence that music in classical Greece was anything other than monotonic – based on a single melody line. So, contrary to Zarlino's suggestion in *The Art of Counterpoint* that an ancient tuning system was oven-ready for modern harmony or counterpart, there was nothing in the historical record to suggest that polyphony and counterpoint had existed in European music before the early fifteenth century. The Greeks, Mei showed, had rejected polyphonic music because they believed that several different melodies and pitches played at the same time would dilute the emotional impact of a piece.

In 1581, Galilei published his *Dialogue on Ancient and Modern Music*, in which he publicised many of Mei's historical insights and threw into question Zarlino's thesis that ancient theory was appropriate for the practice of contemporary music. Galilei started looking at the ways in which the musical instruments of his time were tuned and found that none of them, in practice, kept to the strict numerical relationships described in Greek texts. His entire outlook differed from that of Zarlino in that he was an instrumentalist, a proponent of secular music and, above all, an experimentalist. The specific tuning of instruments, he found, wasn't based on an underlying set of natural or mathematical laws but on the human ear, which simply became used to hearing tones and the intervals between them played in a certain way. According to Galilei, scales, modes and tuning systems were all culturally dependent and had nothing to do with universal or cosmic harmonies.[5]

This was a radical departure from the orthodox musical theory of the day. Galilei argued that the final arbiter of 'good' and 'bad' music wasn't any abstract mathematical scheme but simply *what sounded right*. It was, he insisted, a matter of taste. At the same time, Galilei was no fan of polyphony. On the contrary, he urged that, in this regard at least, we should follow the lead of the ancients and stick to music that had a single melodic line accompanied by a simple orchestration.

In Galilei's view, the power of music didn't lie in mathematical harmonies but in simple melodies that relied on words to stir people's emotions. He and other like-minded thinkers became interested in reviving Greek monophony and drama. Again Galilei consulted Mei, who was spearheading efforts to bring back Greek musical drama through his involvement with the Florentine Camerata, a group of poets, musicians and intellectuals founded by Count Giovanni de' Bardi. The art form that emerged from the efforts of this group became known as recitative – a style of delivery in which the singer adopts the rhythms and delivery of ordinary speech. In conjunction with monophony it served as the basis for opera, the first example of which is widely considered to be *Dafne*, written in 1597 in Florence by the singer and composer Jacopo Peri. During the Baroque era, following the Renaissance, opera would take Europe by storm and become a spectacular, lavish affair full of florid arias and ornate stage sets.[6]

Supporters of the new recitative style scorned polyphonic forms of music, such as the madrigal, in which several voices – typically three to six – sang interweaving melodic lines. Polyphony, of course, survived and flourished after the Renaissance. In fact, Galilei's argument that the human ear should be the final arbiter of taste in music was used against him when it came to the madrigal and other polyphonic forms popular at the time. After all, proponents said, these genres were just as capable of stirring emotions and ennobling their listeners as anything that neo-Greek musical drama had to offer. Thankfully, in the Baroque era, composers didn't have to choose. Recitative and other monodic forms were integrated into early opera, whereas polyphonic forms flourished and evolved into the fugue and concerto.

Through Mei, Galilei learned of the work of Aristoxenus who, eighteen centuries earlier, had opposed tuning methods based on simple number ratios. Mei encouraged Galilei to try tuning two different lutes: one according to the dictates of the theoreticians and the other such that it actually worked well in practice, based on intervals between the notes that were more evenly spaced. Galilei became convinced that 'equal temperament' was the way forward even though it meant abandoning the classical goal of having a perfect consonance for special intervals or, at least, fairly small whole-number ratios between the lengths of string that produced specific note spacings in the scale. As for dissonance, he took a modernist position: allowing dissonance 'if the voices flow smoothly'. In his investigation of tuning and keys, he composed twenty-four groups of dances in 1584 related to twelve major and twelve minor keys.

With the rise of instrumental polyphonic music, it was inevitable – purely from a practical standpoint – that the twelve-tone equal temperament scale would eventually win out. Galilei was among its principal advocates, but not its first and not in the strict mathematical form in which we use it today. Early discussions of it are found in the writings not only of Aristoxenus but also of a Chinese mathematician, He Chengtian, who lived from 370 to 447 CE. The first mention in the West of equal temperament related to the twelfth root of two appeared in a manuscript titled *Vande Spiegheling der Singconst* ('On the Theory of the Art of Singing') by the Flemish mathematician Simon Stevin, although it wasn't published until 1884 – nearly three centuries after his death.[7]

Through his experiments in tuning, Vincenzo Galilei became a pioneer of acoustics research. It was widely assumed at the time that just as the ratio of lengths of two strings with the same tension, tuned an octave apart, is 2:1, the ratio of the tensions needed to produce an octave interval on two strings of equal length would also be 2:1. Galilei put this assumption to the test by hanging weights from strings. He found that, in fact, the ratio of tensions was 4:1. To produce a perfect fifth, the string lengths had to be in the proportion 3:2 but, Galilee found, the tensions in strings of equal length had to be in the ratio 9:4. The ratio of tensions that gave a certain interval wasn't the same as the ratio of lengths, after all. Instead, it varied as the *square* of the string lengths. Galilei had discovered what may have been the first non-linear mathematical description of a phenomenon in nature.

Around 1560 Galilei moved from his home city of Florence to Pisa, where he married Giulia Ammannati. Together they had six or seven children, the eldest of which was Galileo Galilei, eventual discoverer of the craters of the Moon, the four biggest satellites of Jupiter and the law of falling bodies. Galileo, in the spirit of Renaissance science, was an experimenter – a trait he almost certainly acquired from his father. While Vincenzo was conducting his acoustic experiments in the 1580s, with lute strings and weights, it seems likely that Galileo was still living at home, tutoring local students in mathematics. There's good reason to suppose that he may have helped his father investigate how pitch varied with tension and so been exposed to the idea – new at the time – of testing hypotheses by actually going out looking rather than just abstractly theorising. Vincenzo certainly taught Galileo how to play the lute (another of his sons became a virtuoso of the instrument), and it's even possible that this played a part in Galileo's discovery of the law by which objects fall under gravity.

Letting something drop vertically and trying to time the length of its fall is well-nigh impossible without a split-second timing device, which didn't exist in the sixteenth and early seventeenth centuries. Galileo realised he had to dilute the effects of gravity somehow if he were to make progress in understanding how this universal force worked. His ingenious solution: roll balls down a gently inclined plane. Some historical accounts suggest he measured the time to roll to the bottom of the slope using his own pulse. But there's another possibility: taking his cue from the frets on a lute's fingerboard, he may have added moveable frets to his inclined plane, thereby breaking up the ball's continuous motion into discrete intervals of time. As a ball rolled over a fret it would have made an audible click, lending sound as well as sight to the observations.

Galileo also took his father's investigations of vibrating strings further, looking at the effect of cross-sectional area and density of the material used. The results of their combined researches were published in *Discourses and Mathematical Demonstrations Relating to Two New Sciences* in 1638, four years before Galileo died. It presented a whole theory of vibrating strings, pitch, consonance and dissonance – though it was far from the end of the story, as we'll see in Chapter 9.

While Vincenzo Galilei was laying the basis for the introduction of equal temperament in Western musical tuning, and he and his son were founding experimental acoustics, Gioseffo Zarlino was pioneering a revolution of his own. The great theorist was a master of contemporary polyphony and counterpoint. His studies in these areas led him to stress the importance of triads – three notes separated by thirds played simultaneously. Previously, the emphasis was on intervals, or the difference in pitch between just two notes.

Strictly speaking, Zarlino didn't actually talk about 'triads'. He spoke instead of a *harmonia perfetta* ('perfect harmony') when referring to an interval of a fifth that was divided by a third voice into a minor third below the division and a major third above. Yet the idea of a triad was emerging: three notes together forming a standalone harmonic entity.

As the Renaissance drew to a close, composers such as Giovanni Palestrina and William Byrd produced intricate polyphonic music using new techniques to create harmonies and rhythms that had never been heard before. As musicians experimented with multiple, interacting melodic lines, there was a gradual shift from a more horizontal, contrapuntal approach towards progressions of vertically stacked notes. By the early seventeenth century, the first real theories of the triad were introduced, likely due to the standardisation of certain types of vertical progressions that had emerged in the late Renaissance. Only during the eighteenth century did a full theoretical understanding of triads in the modern sense come together.

In the Baroque era and beyond, highly ornamented, intricate pieces by composers such as Bach and Handel were designed to evoke powerful emotions in the listener. These were ambitious pieces. They demanded the development of the orchestra as the dominant musical ensemble. More than that, they created a need for new instruments whose sounds could carry and be heard across much larger spaces.

CHAPTER 7

Instruments of Progress

THE FIRST MUSICAL instruments surely arose by chance, as our prehistoric forebears were awed by the sounds of nature – the whistling and howling of wind, the crashing of waves, the crack of thunder. To these add the rings, taps and other noises made by various materials, tools and weapons as they were used, and there was a broad palette of sonic possibilities to inspire the earliest musical devices. Generation after generation then refined these designs. They iterated, innovated, experimented and tested to produce, over time, the remarkable array of instruments we know today. Percussion instruments based on simple wooden slats evolved into marimbas; metal shields and helmets morphed into gongs and bells; the taut string of a military bow became the strands of a lyre; uniform tubes of bamboo or animal leg bones were flutes-in-waiting. The technology of music became ever more impressive. Hydraulic mechanisms, then electric motors, began to power instruments such as pipe organs. The harnessed feedback oscillations of electric amplifiers gave rise to the limitless possibilities of electronic instruments.

In 1914, two musicologists, Austrian Erich von Hornbostel and German Kurt Sachs, set down a way of classifying musical instruments according to *how* they make sound. The

Hornbostel–Sachs scheme recognises four categories to which a fifth was added later. Aerophones, such as flutes, trumpets and pipe organs, produce sound by vibrating columns of air. Membranophones, which include all types of drums, work because of vibrating membranes. With idiophones, like the xylophone, the very instrument itself vibrates, while chordophones, such as the violin or piano, create sound from quivering strings. More recently, electrophones, such as the theremin and synthesiser, based on electric circuitry, have formed that fifth category.

This is a helpful and strong classification, but then comes the nuance. So many other factors affect the sound coming from an instrument: the material of its construction, its size and shape, and the way it's played. A musician, stringed instrument in hand, may pluck, strike or bow, each technique eliciting a distinctive tonal quality. A wooden instrument struck by a beater sounds very different from a metal instrument hit in the same way, even if the two are identical in every other sense, because their materials vibrate in unique ways. On the other hand, it's hard to tell the difference between the sounds coming from a silver flute and those from a wooden one. In this case the vibrations aren't in the material itself but in the column of air it encloses. The characteristic timbre of wind instruments depends on other factors, such as the length and shape of the tube.

How a musical instrument works is ultimately a matter of physics, and since the laws of physics are the same everywhere, it's not surprising that similar types of instrument have sprung up independently all over the world. A huge variety of drums, flutes and string instruments have been invented, multiple times, in different parts of the globe. All are based on the idea of causing vibrations in some way that are perceived as being musical or rhythmic. How they've developed is influenced by local factors – the kind of materials to hand and the level of technological skill available to work with them, mythic and other beliefs, and regional patterns of travel and trade.

People who live in the frozen north, habitually surrounded by ice and snow, will naturally make instruments from the bone, gut and skin of polar animals. Those who inhabit the tropics will turn to wood, reed and bamboo. Societies that depend on hunting or herding often fashion instruments from the bone and hide of particular animals for spiritual reasons, even when other materials are available. As civilisation dawned, and trade and migration became common, musicians and their instruments travelled far and wide. The sounds they made, the musical styles and the technologies they took with them were part of this artistic diaspora. So, for instance, the design for a certain type of lyre or drum, which originated in one place, would cross hundreds or thousands of miles and be transplanted elsewhere.

In the beginning, musical instruments were based on things found in nature or objects made for another purpose. In 1931, archaeologists came across an 18,000-year-old conch shell in the Marsoulas Cave in the French Pyrenees, famous for its Palaeolithic rock art and ornaments. For eighty years, believed to be a ceremonial drinking cup, it lay forgotten in the Natural History Museum of Toulouse. Then this gastropodal relic caught the eye of archaeologists at the University of Toulouse, who were taking stock of the museum's artefacts. Looking more closely at the shell, they made a surprising discovery: far from being a drinking vessel, the Marsoulas shell had almost certainly been crafted to serve as a wind instrument.

The apex – the pointed tip of the shell – is missing, believed to be due to some ancient accident. But there's no way, according to malacologists (biologists who study molluscs), that the tough, 0.8-centimetre-thick end of a conch shell could have snapped off through wear and tear in the ocean while the rest of the shell remained intact. What's more, computed tomography images revealed subtle evidence of modification by tools: a series of small impact marks in a ring around the opening at the apex where the shell had apparently been struck to split off the tip. The result was a 3.5-centimetre-wide hole at the end, leading into the shell's spiralling inner chambers. The hole would have been the first step in transforming the shell into a wind instrument – the gap through which a player could blow air into the shell. The researchers fed a miniature endoscope into the shell's interior and found another hole in the columella – the central pillar – connecting the broken apex with the spaces inside. Tellingly, the hole bore miniature grooves from the tool that had been used to drill or file it.[1]

Players of shell trumpets today often put their hands into the mouth of the shell to modify the sound coming out. The researchers of the Stone Age conch found that the lip-like edge around its opening had been chipped and shaped, evidently to make it smoother and easier to use. A brownish residue of some substance on the inner and outer surfaces of the apex may have helped hold a mouthpiece in place.

The big question was: could the Marsoulas conch actually produce notes? A musician was invited by the research team to give it a try. He blew into the opening at the end, vibrating his lips as if playing a trumpet or trombone, and, for the first time in eighteen thousand years, brought the instrument to life. Three magnificent, clear tones emerged: roughly a middle C, a C sharp and a D. In a sense, these notes whisk us back in time to the Upper Palaeolithic and the people who would have listened to the same sounds magnified by the walls of their rocky dwelling.[2]

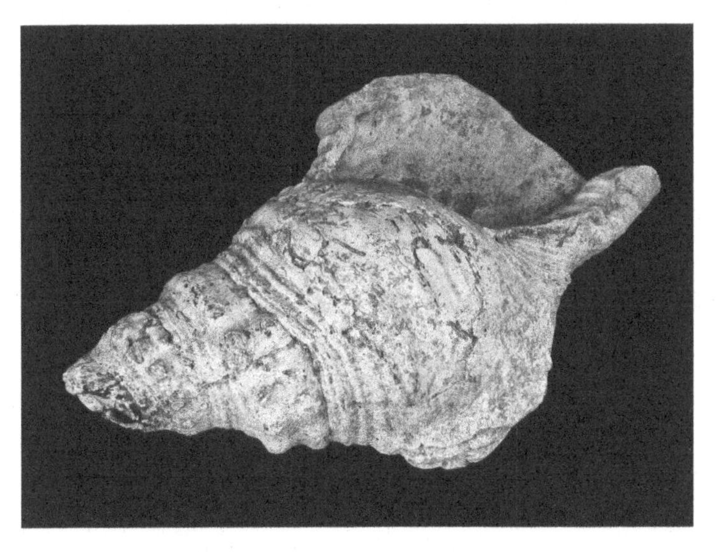

The Marsoulas conch appears to have been purposely altered so that it could be played as a wind instrument.

Wind instruments work by making the air inside them vibrate. Because the cavity of a conch shell takes the form of a spiral, its acoustics are pretty similar to any instrument, such as a French horn, with a chamber that wraps around in a coil. In both cases the initial vibration comes from the lips of the player.

With all wind instruments, a column of air is caused to vibrate at a specific frequency when air is blown either into or across an opening at one end of a tube. In the case of brass instruments and their ilk, a player presses and buzzes their lips against the mouthpiece. Woodwind instruments, on the other hand, come in two varieties. Some, like the clarinet and oboe, are equipped with a reed for making the air column vibrate. Others – members of the flute family – are held sideways so that air can be blown across the end hole to cause the vibration. Notes of different pitch are produced by opening or closing holes along the length of the instrument or, as with the trombone, using a slide, to lengthen or shorten the air column. A skilful player can also play different notes from the harmonic series of a brass instrument (or its relatives) by changing lip tension, or from any wind instrument by a technique called 'overblowing'.

The strings of a stringed instrument produce hardly any sound on their own. They need to be attached to some kind of soundboard, which in turn is part of a resonating chamber like the body of a guitar or violin. As a string oscillates back and forth it causes the board to vibrate, often via a wooden bridge, at the same frequency. Although the moving string would have the same amount of energy with or without the board, the much bigger surface area of the soundboard excites a greater volume of air and so produces an amplified sound.

A key element of most stringed instruments is the sound hole, which can vary greatly in shape and number. An acoustic guitar, for example, generally has a single, circular sound hole, while the lute is equipped with rosettes and the violin with f-holes. While some of the sound from an instrument comes from its vibrating surfaces, a sound hole provides an opening into the resonating chamber, which allows vibrating air inside the chamber to travel outside. The natural vibration mode of the air within the body, known as air resonance, enhances the sound radiation at lower frequencies, producing a deeper and louder overall tone.

Many musical instruments, including those in the violin, lute and oud families, have evolved complex sound-hole geometries through centuries of trial and error. Those of lutes and ouds (short-necked, fretless, lute-like instruments from the Middle East) feature one or more circular holes, often with elaborately carved covers. In 2015, researchers at the Massachusetts Institute of Technology (MIT), in collaboration with a group of violin makers, published a study of how the f-holes of violins had changed over time. They focused particularly on some of the most prized violins in the world – those crafted in the seventeenth and eighteenth centuries in the northern Italian workshops of Antonio Stradivari, Giuseppe Guarneri and the Amati family. Throughout the Renaissance and Baroque eras, these master luthiers fashioned increasingly powerful, richer-sounding instruments, among the most notable features of which were gradually more elongated f-holes and thicker back plates.

The MIT team investigated the effect of changing the shape of the f-hole and concluded that the air flow from inside the body is proportional to the length all the way around the sound hole – its perimeter – and not to its area. Their mathematical analysis showed that the highest air flows through an f-hole occur at the places at the top and bottom where the opening is narrowest. The effect is similar to what happens if you put your thumb over the end of a hose to speed up the jet of water coming out.[3]

Of course, the reason a Stradivarius violin costs a fortune today is *not* the shape of its sound holes. Makers of fine string instruments are very picky about the materials they use in construction. They select certain types of wood, called tonewoods, known to have specific sound qualities. A guitar maker might use a combination of spruce and mahogany; for violins, the premium woods are spruce and maple. And for the best instruments, these woods are sourced with meticulous care.

In 2011, a Stradivarius sold at auction in New York for a record $15.9 million. Made in 1716, it's nicknamed the Lady Blunt after Lady Anne Blunt, the daughter of Lord Byron, who bought it in 1822. A similar instrument, crafted in the same year, known as the Messiah, resides in the Ashmolean Museum, Oxford, and has been valued at $20 million. Such astronomical sums reflect not just the exquisite craftsmanship of these violins and their almost perfect state of preservation, but also the unique acoustic quality of the woods used to make them. Luthiers have long sought to learn the secrets of Stradivari and the other great Italian instrument makers of this time from Cremona. In recent years, scientists have shed some light on what may be one of the key ingredients.

From about the sixteenth to the nineteenth centuries, and possibly earlier, the world went through a cold spell. Global temperatures dipped below normal but especially so in Western Europe and regions around the North Atlantic. It was during this period, known as the Little Ice Age, that the spruce and maple trees used to make Stradivarius violins grew. The drop in temperature, which coincided with a period of reduced solar output, hampered the normal growth of the trees. It meant there was less difference between the lighter, spongier wood laid down during spring and summer months and the denser wood produced during periods of slower growth in autumn and winter. Sound travels more freely through high-density wood, so the exceptional acoustic properties, including the intense resonance, of eighteenth-century Cremonese violins could be at least partly explained by climatic conditions prior to their construction![4] An astronomer interested in music might point out a cosmic connection. The gorgeous, emotive sounds of a Stradivari played by, say, Itzhak Perlman or Anne-Sophie Mutter, are at least partly indebted to the behaviour of our neighbourhood star.

Other factors were undoubtedly at work in the creation of these unique instruments. Specialist craftspeople often keep closely guarded secrets, which in the case of premier luthiers involve the various chemicals they apply to the wood and the varnish they use to seal it. Analyses of Stradivari maples reveal that they were treated with substances containing minerals such as aluminium, calcium and copper. The wood was further transformed by age and vibration while playing into a complex composite material virtually impossible to replicate by modern violin makers.

Within seconds, you'd be able to tell the difference between an expertly made violin and an inexpensive one intended for a beginner. Even if they were playing the same note, you could easily tell them apart. The difference in sound would be much greater between a violin and, say, a clarinet, or a cello and a bassoon. But what's the basis for these differences?

The moving parts of musical instruments, such as strings, vibrating columns of air or drumheads, oscillate back and forth in complex ways. Not only do they produce some definite, intended note – the fundamental – but also a host of other notes at different frequencies, known as overtones. Each of these contributions to the overall sound – the fundamental and the overtones – is known as a partial.

An overtone that's a whole-number multiple of the fundamental is called a harmonic overtone or, simply, a harmonic. Confusingly, the first harmonic is the fundamental, the second harmonic is the first overtone and so on. Play the note G2 (the G two octaves below middle C), which has a frequency of 98 Hz, on a cello, and mixed in with the sound of that root note will be the second harmonic at double the frequency (196 Hz), the third harmonic at three times the fundamental frequency (294 Hz) and so on. The note of A4, with a frequency of 440 Hz, played on a flute, will be accompanied by harmonics at 880 Hz, 1760 Hz and other whole-number multiples of the fundamental. This fact alone doesn't explain the difference in sound quality, or timbre, between different instruments. They may be playing the same root note, but the relative level or intensity of each harmonic will differ between two different instruments, such as a clarinet or flute. A graph of the relative intensities of the harmonic overtones is known as an instrument's frequency spectrum.

Even if two instruments produce exactly the same overtones with similar amplitudes, they can sound quite distinct. That's because the relative amplitudes of their overtones may change *over time* in different ways. The variation in amplitude of an overtone over time is known as its time envelope. The combination of time envelopes of all the different overtones is a vital factor in determining the tonal quality of a specific instrument.

In addition to harmonic overtones – tones that are whole-number multiples of the fundamental – instruments may give rise to inharmonic overtones. These don't have any whole-number relationship with the fundamental and therefore don't reinforce the fundamental in the way that harmonic overtones do. They give rise to a less musical, 'noisier' sound.

Drums are a classic example of instruments that have a strong fundamental but a highly inharmonic overtone sequence. Unlike a string or an air column, a drumhead doesn't vibrate simply in a single direction. A vibration initially spreads out from where the membrane of the drum is struck and is then reflected at the outer edge, like water waves in a circular bowl. The way the various parts of the surface undulate after that is complex and calls for detailed mathematical analysis. Suffice it to say that instead of giving overtones that are two, three, four or other integer multiples of the fundamental, a drum might produce overtones at 1.59, 2.14, 2.30 and so on, times the frequency of the root note. Most drums, consequently, are indefinitely pitched.

A drum can be tensioned to produce a specific fundamental, say at 220 Hz, but it won't sound like this same note (the A below middle C) played on a piano or guitar. The perception of a certain pitch is lost amid the cacophony of inharmonic overtones. It's possible, however, to reduce or mute certain overtones when designing and playing drums so that the resulting overtone sequence better approximates a harmonic ideal. Such is the case with timpani, or kettledrums, which can be tuned by adjusting (usually eight) tension rods spaced at equal intervals around the circular edge of the drum. Players aim to have an even tension over the entire drumhead and then to strike the drum in certain places. Striking the head about one-third of the radius from the edge gives a strong fundamental plus a series of overtones that are close to the harmonic overtones we're used to hearing from other instruments. If the mallet comes down in the centre, on the other hand, the harmonic tones ring out clearer than the fundamental and the characteristic sustained bass of the timpani is lost.

String and most wind instruments, along with the human voice, have a full complement of harmonic overtones. If we call the fundamental frequency f, then a string, or a vibrating column of air open at both ends, will give rise to overtones at $2f$, $3f$, $4f$, etc. An air column that's closed at one end, as in the case of an organ pipe, panpipe or clarinet (the last because of the nature of its mouthpiece), will produce only odd harmonics: $3f$, $5f$, $7f$, etc., in addition to the fundamental. The situation is more complicated with other sources of vibration in music, as we've seen in the case of drums. Do the maths behind the physics of a metal bar that's free to vibrate at both ends, for example, and you'll find that in addition to the fundamental there are inharmonic overtones at $2.76f$, $5.40f$, $8.93f$ and so on. This blend of specific frequencies is what gives tubular bells, glockenspiels and other types of chime their specific sound.

In the case of the human voice, only vowels and other sounds (like 'mmm') that can be sustained have a definite pitch; most consonants, particularly plosives or popping sounds, such as 'b' and 'p', are unpitched. When the voice of a trained singer is used as a musical instrument, the vocal tract can be adjusted to vary the amplitudes of the overtones, called formants, which define different vowels.

There are subtleties, too, in the sound of other instruments depending on how they're played. A string that's bowed, as on a violin or cello, produces a sound wave with sawtooth qualities as the string is repeatedly grabbed by the rosin on the hairs of the bow and then released. A brass instrument will have some pulse-like qualities as air is periodically forced through the lips into the mouthpiece. Vibrato (slight variations in pitch that create a wavering effect), and the attack and decay of notes (how the notes begin and end), also contribute to an instrument's distinctive timbre.

Musical instruments have evolved over time as the technology available has progressed and as music itself has changed and placed new demands on instrument makers. This progression has been most obvious in Europe, especially since the beginning of the seventeenth century and the dawn of the Baroque era. But well before then, going back to ancient times, ingenious inventors started coming up with – and continue to discover to this day – new ways of creating musical sounds.

The earliest ancestor of modern keyboard instruments was the hydraulis, a type of pipe organ invented in Greece in the third century BCE. In this fantastic piece of antique engineering, water enters through a pipe from above and is mixed with air from a side pipe. The air is compressed as the water descends. Water and air then arrive together in the *camera aeolis*, or 'wind chamber', before being separated. The compressed air is temporarily stored in a wind chest after which it's blown through the organ pipes to produce sounds. It didn't take much effort to create an impressive effect, as the fourth-century Latin poet Claudian made clear: 'Let him thunder forth as he presses out mighty roarings with a light touch.'

One of the most spectacular of water organs is sited at the Palazzo del Quirinale, a magnificent building completed in Rome in 1583 that's served as the residence for thirty popes, four kings of Italy and twelve presidents of the Italian Republic. The present organ, however, is a recent addition.

It's located in the gardens of the palace and was built in the late 1990s following a design similar to that of a nineteenth-century organ. Water from a hilltop spring courses into a stabilising room – a temporary reservoir – in the palace where its pressure can be kept relatively constant. It then drops some eighteen metres into the wind chamber resulting in enough air pressure to power the instrument, which is played from a single keyboard of forty-one notes.

The familiar church pipe organ, driven by leather bellows, first appeared around 900 CE and by the fourteenth century was a common fixture in cathedrals around Europe. A much smaller keyboard instrument, the clavichord, was invented in the fourteenth century but used mainly for practice and composition. Its sound was generated by small metal blades, called tangents, striking brass or iron strings whose vibrations were transmitted by a bridge to the soundboard. Although it was too quiet for large performances, its popularity continued for several centuries thanks to its expressive tone, portability and comparatively low cost. It was the first choice for anyone wanting to learn keyboard and was popular with composers, such as Mozart, as a travel instrument when on tour. The harpsichord also first emerged in the late Middle Ages but was larger and worked in a different way. Its strings weren't struck but instead were plucked by small plectrums made from quill. The harpsichord's distinctive sonority identifies it immediately with Renaissance and Baroque music. But it's limited by the fact that the player has no control over the loudness of its tones. No matter how hard or softly you press the keys, the sound will always have the same volume.

Out of the quest for a more dynamic keyboard instrument capable of a greater range of sounds, in both amplitude and expression, emerged the piano. Its earliest incarnation came in 1698 at the hands of Italian instrument maker Bartolomeo Cristofori, who called his creation *gravicèmbalo con piano e forte* ('harpsichord with soft and loud'). Happily, this was quickly shortened to *pianoforte*, the name reflecting the fact that the player could control the volume by adjusting the force with which each key was struck. Just as the unique sound of the harpsichord is characteristic of the Baroque period, the more forceful presence of the piano marks it as the sonic standard-bearer for the Classical and Romantic periods.

Early pianos used a combination of brass strings (for the lowest notes) and iron strings. As time went on, composers such as Mozart, Beethoven and Liszt demanded more and more of the instrument, especially in terms of range and volume, so that the sound could fill ever-expanding concert halls and be heard above an orchestra. In response, designers added more strings and increased their tension, which meant they had to strengthen the frame with the addition of an iron plate. The biggest change began in the mid-nineteenth century when steel strings were introduced. By 1912, these had reached their modern form with a tensile strength three times greater than that of the iron wires in the earliest pianos. Astonishingly, the combined tension of all the strings on a concert grand piano can exceed twenty metric tons!

Other instruments also evolved over time. Beginning in the seventeenth century with the dawn of the Baroque period in music, composers began writing more elaborate and emotionally charged works, adding more parts to complement or replace the human voice. Many existing instruments that had limited ranges and dynamics were now seen as insufficiently expressive and therefore fell from favour. The shawm, a double-reed instrument and ancestor of the oboe, had been popular throughout the Middle Ages and into the Renaissance. But although loud, and therefore mainly used by town bands and other outdoor ensembles, it was seen as too inflexible and shrill for the more complex, polyphonic music that evolved at the start of the seventeenth century. Nevertheless, the shawm survives to this day in the folk music of some regions of the world, having also gradually evolved into other forms. Musicologists had been puzzled by historical references to 'still shawms' that were said to have had a more mellow sound. Then, in 1980, one of these instruments was found aboard the wreck of the *Mary Rose*, a ship of Henry VIII's navy, which sank in 1545. A reproduction of this sole-surviving still shawm was made and, when played, found to have a deep, rich tone. Supported by fiddles and tabor (a handheld drum), remains of which were also recovered from the wreck, it would have provided the accompaniment to dancing aboard the vessel.

A host of other mediaeval and Renaissance instruments, with such evocative names as 'crumhorn', 'flageolet', 'sackbut', 'citole' and 'psaltery', evolved into new forms or fell by the wayside as musical tastes changed. Horn instruments were originally just that – animal horns into which players blew while vibrating their lips. In time, natural horn was replaced by metal or other materials. In the mid-seventeenth century the hunter's horn began a transformation that involved a lengthened tube, a narrowed bore, a widened bell and, in consequence, a much wider range. By 1725, the metamorphosis was complete, and what we now know as the French horn had emerged.

From the fifteenth to the seventeenth centuries, one of the most popular and versatile instruments was the cornetto. Consisting of two hollow, gently curved pieces of wood, wrapped in leather and pierced with six fingerholes and a thumbhole, it was played by buzzing into an acorn-shaped mouthpiece. In its heyday, it was a rival of the violin, prized for the resemblance of its sound to a human voice and ability to perform elaborate ornamentation. In fact, many works of the early seventeenth century were marked 'violino o cornetto' (violin or cornetto) implying that either was a suitable choice.[5]

But the cornetto was fiendishly difficult to play at a level that could match its great string rival. The violin also had the advantage that it could sustain a note indefinitely – no breath required! Even at its height, there were relatively few virtuoso players, such as Giovanni Bassano and Girolamo Dalla Casa, both of whom worked in Venice, which was also a major centre for both the teaching and manufacturing of the instrument. Then disaster struck. Many of the leading players and teachers fell victim to the great Venetian plague of 1630. By the second half of the seventeenth century, the cornetto was in sharp decline as violin technique developed and other wind instruments, such as the oboe, became more sophisticated.

As the middle class expanded in eighteenth- and nineteenth-century Europe, so did the demand for more sophisticated entertainment. What was once available only to the aristocracy now became accessible to the *nouveau riche*. At the same time, composers demanded more and more of instruments in terms of range, timbre, expressive ability and volume. During the Classical and Romantic periods of music, from about 1750 to 1900, scores were produced for increasingly large orchestras, and the growth in size of audiences and concert venues meant that some instruments had to become louder to be heard at the back of big halls and amid a sizable ensemble. Instruments that had been easy to hear in antechambers and private ballrooms now needed to be audible throughout much larger spaces. Metallic flutes superseded those made of wood, and in some cases steel replaced gut for strings.

The range and flexibility of instruments also were broadened. The early nineteenth century saw a revolution in the manufacture of brass instruments: the addition of tubing of different lengths through which air could be directed by the depression of valves. No longer were the players of horns and trumpets limited to the notes of the natural harmonic series – they could roam across the entire twelve notes of the chromatic scale. New instruments, such as the clarinet, saxophone and tuba, became fixtures in orchestras. In some cases various subspecies sprang up: small clarinets, standard clarinets, bass clarinets and so on. The proliferation continued into the twentieth century as technology progressed. Electronic instruments, which we'll explore in Chapter 10, also burst onto the scene.

But before we move forward into the present day and the exotica of modern music, we must again take a step back into the past. As various instruments increasingly began to be played together, the issue of accurate pitch became crucial. After all, who would want to listen to an orchestra that was audibly and obviously out of tune?

CHAPTER 8

Pitch in Time

IF YOU'RE SINGING on your own, it doesn't matter where you start. So long as the rest of the song follows that opening note, you should sound all right (if not quite ready to debut at La Scala). Similarly, next time you're at a birthday party, listen carefully as everyone launches into 'Happy Birthday'. While the first few notes will slide all over the place, within a line or two, most slip into a communal tuning – a melodic compromise that suits the whole group.

Absolute or 'perfect' pitch – the ability to reel off any note on the scale at its specific frequency – isn't necessary in situations like these; nor is it important if you're accompanying an instrument, like a guitar, providing you adjust your voice to match the instrument's tuning. For thousands of years, as far back as prehistory, musicians didn't worry about perfect pitch. Relative pitch was all that was required – and all that was available.

Imagine you're living in Athens circa 400 BCE. A couple of your more musical friends are visiting for an old-school symposium, followed by an informal jam session. Calliope, who also sings, is bringing her lyre and Apollodoros his aulos – a reed instrument with two pipes connected to a single mouthpiece. After a couple of glasses of local wine, you start tuning up. The notes Apollodoros can play on his aulos are determined by set spacings between the fingerholes. Unless he has a drill to hand, he can't change these, so you and Calliope must tune your lyres to match, tightening or slackening the strings, respectively. There's no other option. Tuning forks won't exist for another two thousand years, and this is long before instruments were mass-produced to a particular standard pitch.

Remember, there's an important difference between pitch and frequency, even though these terms are often used synonymously. Frequency is an objective, physical quantity – the number of vibrations of a wave per second, which nowadays can be measured precisely. All you need is a mic and an oscilloscope. The mic picks up the sound and turns it into an electrical signal, which is then displayed on the oscilloscope as a wave with a specific frequency. Pitch, on the other hand, is a perceptual phenomenon that depends on the listener and therefore can't be measured. Despite these differences, there's a close relationship between the two. People generally agree that the greater the frequency of a note, the 'higher' it sounds.

Back in your imaginary Athenian home, Apollodoros blows a note on his aulos; you and Calliope adjust the strings on your lyres accordingly. The intervals between notes are well defined in the ancient Greek musical system, so once you've got one string tuned in, the rest can be easily tweaked – you're good to go.

This is how people managed to play instruments and sing together harmoniously in the many centuries before standardised pitch. They used their ears and tuned locally to the most convenient reference point. Notes were defined in relation to each other according to standard scales or modes, and instruments were tuned relative to each other following the prescribed pattern.[1]

Some instruments are seriously difficult to retune once they've been made. An obvious example is the church organ – one of the main ways of accompanying the human voice in sacred music in Europe throughout the Middle Ages. A pipe organ is a gigantic thing – immovable and with no means of altering its tuning except by lengthening or shortening dozens of metal pipes. Fortunately, the voice is much more flexible. But here's the thing: because there was no agreed standard for tuning back then, different organs in different locations, playing the same note of the scale – say, middle C – might produce a sound at very different frequencies. If these organs had ever been brought together and played at the same time, they'd have sounded awful.

Pitches didn't just vary from one town to another but could differ wildly even within the same place. The pitch of a note on an English cathedral organ in the seventeenth century might be as much as five semitones, or a perfect fourth, lower than the same note played on a domestic keyboard instrument half a mile away. A church organ could also change in pitch over time because of wear and tear. The end of an organ pipe could be hammered inward or flared outward by a cone-shaped tool to sharpen or flatten the note it made. Over time, after repeated adjustments like this, the pipe ends became frayed and all had to be trimmed down, thereby raising the organ's overall pitch.[2]

Before the Renaissance, no one gave much thought to a standardised system of tuning. Most serious music, both sacred and secular, was sung, and all it took was for someone with a good sense of pitch to give the starting note. Where instruments were involved, they were few in number and everyone quickly tuned to whatever instrument was the least capable of having its pitch adjusted. Even into the sixteenth century, a common practice was referred to as 'pitch of convenience'. In particular, for an a cappella (literally, 'chapel style') performance, the important thing was, as the composer and musical theorist Ludovico Zacconi wrote in his *Prattica di Musica*, published in 1596, 'to have regard for those who are to sing, that they be at ease with the pitch, neither too high nor too low'. In other words, nothing specific.

Problems with tuning started to mount during the Renaissance, when it became increasingly common for groups of musicians to play different kinds of instruments together. Performing in an ensemble meant dealing with several different working pitches. For strings the solution was quick and easy: retune them on the spot. But a wind player might have to purchase completely new instruments when moving from place to place and venue to venue just to be able to comply with local tuning preferences.

This sort of thing obviously couldn't go on forever. As the Baroque era dawned, in the mid-seventeenth century, efforts at pitch standardisation stepped up a gear. The Hotteterre family, renowned Parisian instrument makers, remodelled their entire range of woodwind instruments so that the A above middle C was tuned to that same note played on the cathedral organs in the French capital. These organs, in turn, had adopted the pitch convention, which started in Germany in the early Baroque era, of tuning to so-called *Kammerton*, or 'chamber pitch'. Recognising that musicians would still have to cope with a variety of local pitch preferences, however, the Hotteterres designed their instruments to be adaptable. Their transverse flutes came in sections so that the length of the air column could be made longer to drop the pitch or shorter to raise it. An adjustable plug in the head section was used to correct the tuning and playing properties of the flute if the middle sections were changed. Brass instruments were also made with extra crooks – small lengths of tubing called *corps de réchange* – that could be fitted to the instruments to change their pitch.[3]

In naming pitches today, musicians have a choice of two different systems. One was developed in the 1860s by German physicist Hermann von Helmholtz and is based on how German organ builders labelled their pipes, which in turn derived from mediaeval German organ tablature. It employs a combination of upper- and lowercase letters (A to G), and sub- and super-prime symbols, to denote each note of the scale. The other system, known as scientific pitch notation, which we'll use here, combines the note name (with a sharp or flat if needed) and a number identifying the pitch's octave. In this system, the lowest note on a standard 88-key piano is A0 and the highest is C8. Today, we can equate each of the pitches in this notation with a specific frequency, because there's an international system of tuning in which A4 – the A above middle C – is defined as having a frequency of 440 cycles per second, or hertz (Hz).

The A4 = 440 Hz standard is a fairly recent development. Before that, lots of different conventions had been tried at different places and times. During the Renaissance, when instrumental music first became prominent, relative to vocal music, there was a tendency to gradually pitch notes higher. An instrument that's tuned higher sounds brighter and can be heard more clearly, especially in a large room. But this raising of pitch was a nuisance for vocalists. At the beginning of the seventeenth century, German composer and organist Michael Praetorius reported in his encyclopaedic *Syntagma musicum* that pitch levels had become so high that singers were experiencing severe throat strain and lutenists and viol players were complaining of snapped strings. Various solutions were tried, generally involving the adoption of different pitch standards for organ and voice: *Kammerton* for instrumentalists and *Chorton* ('choir tone') for singers. When voices and instruments had to perform together, as in a cantata, the two sections would often use music written in different keys to compensate for the different pitch standards.

Throughout the Renaissance, then the Baroque period, and finally the Classical and Romantic eras, pitch standards came and went, and different ones coexisted throughout Europe. Because there was no way of measuring or recording frequencies from pre-Classical periods, musicologists still debate exactly what standards were used. Even where we have surviving instruments, there often isn't a clear indication of how they were played – for example, in the case of a wind instrument, how hard they were blown – and therefore of the pitch they produced.

Contemporary musicians who specialise in playing pieces from bygone eras often use modern instruments that have been designed and built according to certain presumed standards of the past. A common assumption is that Baroque woodwind and keyboard instruments were tuned such that A4 sounded at 415 Hz – so-called Baroque 'low pitch'. But there's little historical justification for this precise frequency. The choice is mainly a matter of convenience, because it's almost exactly one semitone below today's standard of A4 = 440 Hz. This makes it handy, for instance, for transposing keyboards built into some present-day reproduction harpsichords. Other historical tuning standards widely used by ensembles specialising in early music range from as low as A4 = 392 Hz, for French Baroque repertoire, up to A4 = 466 Hz, for Italian Renaissance pieces by the likes of Giovanni Pierluigi da Palestrina and Claudio Monteverdi.[4]

In 1711, at about the same time that Bartolomeo Cristofori built the first piano in Italy, the tuning fork was invented in England by John Shore, royal trumpeter in the band of Queen Anne. Shore was famous for his playing, and George Handel, who was then court composer, wrote many of his more elaborate trumpet parts for him. Unfortunately, at a concert that included one of these ornate passages, Shore split his lip so badly that he was 'ever unable to perform'. Instead he took up the lute, and it was for this instrument that he devised the tuning fork. A man of habits, he announced before every performance, 'I never go anywhere without my pitch fork', and then went through the novel procedure of tuning up while the audience was forced to look on.[5]

Shore gave Handel one of his forks, which is still in existence and sounds the note C4 at 512 Hz, equivalent to A4 at 422.5 Hz. It was used at the first performance of Handel's most famous oratorio, *Messiah*, in 1742. Musicologists can also place the exact pitches at which Mozart, Beethoven and their contemporaries of the Classical era, from the 1750s to the early 1820s, intended their works to be heard – and there's quite a variety. Mozart had his pianos and clavichords tuned to A4 = 421.6 Hz, whereas Beethoven's tuning fork produces a markedly different A4 at 455.4 Hz. This last fact might explain why, when rehearsing for the premiere of Beethoven's Ninth Symphony, singers in the chorus complained that the soprano parts were too high. Following the first performance of his *Kreutzer Sonata*, at which he himself played piano, Beethoven gave his fork to the accompanying violinist, George Bridgetower. It later came into the possession of composers Gustav Holst and Ralph Vaughan Williams, and, finally, the British Library, as a gift from Williams's widow in 1992.

Until the advent of electronic meters, the tuning fork was the most trustworthy pitch carrier available – vastly superior to the only previous, purpose-made tuning instrument, the pitch pipe, about which the French philosopher Jean-Jacques Rousseau, writing in 1764, noted 'the impossibility of being certain of the same sound in two places at the same time'. Tuning forks removed that uncertainty and, thanks to the many surviving examples stretching back to the early eighteenth century, enable us to know at exactly what pitch the musicians and composers who used them were working.

A modern tuning fork with sheet music.

You might suppose that a well-made tuning fork would stay 'in tune' at all times – after all, that's their one and only job. But changes in temperature affect their performance: heat flattens them and cold sharpens them, which is exactly the opposite of how organ pipes respond. Rust, too, will flatten them slightly, though modern forks are treated so that this never happens.

Before tuning forks appeared on the scene, one of the few reasonably reliable ways to gauge the pitch standard in use at the time was to look at old church organs. The dimensions of their pipes served as a fairly good guide to the frequency of the notes they played. The evidence of eighteenth- and nineteenth-century tuning forks, though, pinpoints the evolution of standard A4 – concert pitch – much more accurately.

The tuning fork used with the pipe organ in Versailles Chapel from 1795 vibrates at 390 Hz. An 1810 Paris Opera fork sounds an A4 at 423 Hz, the same as an 1815 fork in use at the Dresden opera house. Just eleven years later, however, the same Dresden company had upped its pitch to 435 Hz, whereas by the mid-nineteenth century the orchestra at La Scala in Milán had moved to a concert pitch of 451 Hz.

Different regions, countries and even instrument manufacturers adopted their own standards, making orchestral collaboration a challenge. Despite the efforts of two organised international summits attended by noted composers, such as Hector Berlioz, no standard could be agreed upon.

The frequencies at which historical tuning forks sound, and the way that instruments were constructed, reveal a gradual rise in average concert pitch throughout the Classical era and into the Romantic era. After about 1760, the conventional pitch of A4 climbed from a low of 377 Hz to a high of 457

Hz in 1880 in Vienna, and it's no coincidence that this pitch augment came during a time when orchestras grew in size and large orchestral compositions became ever more popular.

Although in Germany a pitch standard of 440 Hz was recommended by the acoustician and musicologist Johann Scheibler as early as 1834, it was largely ignored at the time. In England the mid-nineteenth century saw the adoption of 'Old Philharmonic Pitch' at about 453 Hz. Again, this proved to be a problem for vocalists, and a compromise pitch at A4 = 435 Hz, known as diapason normal or (in the United States) 'French pitch', was agreed upon by an international commission in Paris in 1858. Nevertheless, a confusing variety of pitches continued through to the end of the nineteenth century. This was reflected in the way pianos were tuned. The English manufacturer Broadwood Piano Co. offered to tune their instruments in three different settings – low (433 Hz), medium (435 Hz) and high (454 Hz) – to suit the various needs of singers, non-professional players and concert musicians. In 1879, Steinway, keen to keep as many of its customers happy as possible, tuned its pianos to A4 = 457.2 Hz in New York and 454.7 Hz in London.

In February 1859, the French government, acting on the advice of a number of distinguished composers, including Daniel Auber, Giacomo Meyerbeer and Gioachino Rossini, passed a law making diapason normal the nation's standard pitch. At the same time, Secretan, a Paris-based company that manufactured scientific instruments, produced an official French Commission Diapason Normal tuning fork to be held at the Paris Conservatory of Music, along with a dozen copies. Other countries gradually adopted the same standard – but not Britain. Not until January 1885 was it officially announced that Queen Victoria had sanctioned the adoption of diapason normal for her private band and that it would in future be used at state concerts. Still, there was no universal consensus in Britain or elsewhere. The Royal Philharmonic Society adopted diapason normal in 1896, but brass bands, both military and civilian, weren't convinced and, in fact, their pitch of choice tended to drift higher over time.

Despite the occasional outliers, though, orchestras, composers and nations across Europe and North America were gradually converging on an accepted standard for concert pitch. In Britain, in 1896, the 'New Philharmonic Pitch' was set at A4 = 439 Hz. Finally, the frequency of 440 Hz, first proposed by Schreiber in 1834, won broad approval. Strangely enough, standard concert pitch tuning of A4 = 440 Hz (A440) was stipulated in the Treaty of Versailles – the 1919 peace agreement that followed World War I. Twenty years later, on the eve of World War II, it was recommended again at an international conference and formally adopted as a standard by the International Organization for Standardization in 1955 as ISO 16.

A frequency difference of five cycles per second separates today's A440 benchmark from diapason normal. In fact, the new standard was initially set at A4 = 439 Hz, but this was criticised on a couple of grounds. First, 439 is a prime number – a number that divides only by itself and one. This made it difficult, at the time, to reproduce in the laboratory. More importantly, though, the issue of temperature hadn't been dealt with properly in the original diapason normal specification. The standard fork held in Paris was stated to vibrate at 439 Hz at 15 °C. The problem is that wind instruments, especially organs, rise in pitch almost in proportion to the increase in temperature of the surrounding air. It was suggested later that in establishing the diapason normal, the French Commission should have chosen a temperature of 20 °C.

Environmental conditions, especially temperature and, to some extent, humidity, are one of the problems that musicians face when trying to keep their instruments in tune. If the ambient temperature rises, the pitch of stringed instruments, like the violin, drops, whereas that of wind and brass instruments rises. Played together, the two groups move in opposite directions and, during a performance, what might start out well enough could become increasingly strained. This used to be more of a problem in the past because, in small concert halls or opera houses, with a crowd of people present, the air would tend to become warmer (and more humid) as the performance progressed. Today, it's less of an issue as auditoriums are usually air-conditioned and instruments, once notorious for their pitch and tuning instability, have become much less affected by changes in their environment. The modern piano, for example, with its full metal frame, is much more stable than its half-metal, half-wood ancestor of a century or more ago.

Concertgoers will be familiar with the sound of an orchestra tuning up before the performance begins. The process starts, most commonly, with the oboe sounding the single note of concert A and then everyone else, as it were, 'pitching in'. When the first orchestras became part of the musical scene in the late seventeenth century, they consisted mainly of string instruments, sometimes augmented by a couple of oboes to strengthen the first and second violin parts. Soon composers were writing separate parts for the oboe, exploiting its singing tone as a contrast to the violins.

The bright, rather penetrating sound of the oboe was easy to hear, and its pitch was more stable than that of gut strings, so it was natural to rely on it for tuning. Other instruments drifted in and out of the orchestra – flutes, bassoons, French horns, clarinets – before its instrumentation became relatively standardised as we know it today. But oboes were almost always present, so it became the standard instrument for tuning.

In 1964, Chicago-based Peterson Electro-Musical Products introduced the first commercial handheld electronic tuner for musicians. Today, such devices are used worldwide – including by oboists. Oboes can play sharp or flat, just like any other instrument, but every oboist uses a pocket electronic meter to ensure that their A is exactly right. In fact, some orchestras use an electronic device to sound the tuning note, though, even then, tradition is respected and it's the oboist who turns the device on for tuning and off when the task is completed!

When a keyboard like a piano is present, it takes precedence. Most audiences might get fidgety in the hour or two it would take a piano tuner to check the pitch of all the strings. So the piano starts by playing an A4 and the orchestra must follow suit. But still the oboe plays a role, being the first to match the note from the keyboard for all the other players to copy. If the performance involves only strings, then it falls on the concertmaster – the lead violinist – to pretune his or her A string to concert pitch and play the note for everyone else to hear.

Orchestras always tune to A. Not only is standard pitch specified in terms of A4 but also every string instrument has an A string. On the other hand, if you happen to play in a concert band, you'll know that the tuning of choice is B flat because most brass instruments are pitched in that note. And just because A = 440 Hz is now a worldwide standard for pitch doesn't mean that everyone has to follow it. Some orchestras favour a slightly higher pitch, which some believe results in a brighter sound. The New York Philharmonic, for instance, tunes to A442 and some European orchestras to A443. The tuning of some well-known popular music is especially interesting – and controversial – as we'll discover in Chapter 11.

CHAPTER 9

Sound Science

FROM THE ACCIDENTAL melodious and rhythmic sounds of nature to the earliest tuneful utterings of the human voice, physics and music have been interwoven. Every musical sound, and every experience of every sound, begins with a vibration and a pattern of waves. Out of that simple fact stems the science of musical acoustics.

From the time of Plato, harmony was considered a fundamental branch of physics, or natural philosophy, as it was known then. More than that, though, it was believed to form an essential link between the individual and the cosmos. Early Indian and Chinese theorists took similar approaches: all sought to show that the mathematical laws of harmonics and rhythms were fundamental not only to our understanding of the world but also to our well-being. Confucius, like Pythagoras, regarded the small whole numbers – one, two, three, four, etc. – as the source of all perfection.

It's surprising how much the Greeks had managed to grasp about the nature of sound more than two thousand years ago. Aristotle taught that sound is a 'movement of alteration'. In Book XI of his *Problems*, he wrote: 'the air which travels makes the sound; and just as that which first sets the air in motion causes the sound, so the air in its turn must do likewise and be partly a motive power and partly itself set in motion'.

Strato of Lampsacus, who lived a bit later than Aristotle, around the turn of the third century BCE, went further. In his book *On Things Heard*, he explained that sound is the result of many separate beats in a medium, such as air, and that its pitch is governed by how often the beats occur. Exactly how he reached this conclusion we don't know – perhaps it was just a lucky guess.

Turning from the theoretical to the practical, the Roman writer Vitruvius, in his ten-volume *De architectura*, addressed the subject of acoustics in theatres. In a chapter titled 'The Echeia, Sounding Vessels in Theatres', he talks about how the Greeks made bronze jars – *echeia* ('echoers') – and placed them in chambers under the rows of seats in amphitheatres. They were custom-designed for each theatre to produce harmonics and, according to Vitruvius: 'They should be set upside down, and be supported on the side facing the stage by wedges not less than half a foot high.' The idea popped up again in the Middle Ages when acoustic jars made of pottery were strategically placed in the walls and floors of mediaeval churches to enhance the sound quality.[1]

The effectiveness of *echeia* has been questioned by modern acoustic engineers but not the architecture of Greek amphitheatres. These structures are both visually stunning and acoustically sophisticated. Everything from the radius of the semicircular seating to the spacing between seats was carefully planned to achieve the optimum sonic effect. Many Greek cities were built on or near hills, so it made sense to carve the local theatre into a natural slope, which would then have a fabulous view to add to its impressive sound qualities. Our word 'theatre', in fact, comes from *theatron*, which means 'seeing place'.

Today, voices and instruments can be amplified at will with the help of microphones and public-address systems. Two thousand years or more ago, the actors and musicians on stage relied on the ingenuity of the theatre shape and construction to project whatever sounds they made into the audience. Modern tour guides will often drop a pin or tear a piece of paper on the stages of these time-worn auditoriums to demonstrate how even the gentlest of noises will carry high up into the seats at the back. Sound too good to be true?

In 2017, a team of researchers from Eindhoven University of Technology in the Netherlands made a close study of the acoustic qualities of three of the greatest surviving Greek amphitheatres: the theatre at Epidaurus, the theatre of Argos and the Odeon of Herodes Atticus, which date from 400 BCE, 200 BCE and 200 CE, respectively. The scientists made lab recordings of various sounds – actors speaking, a dropped coin, a struck match and so on. Then they set up loudspeakers in the centre of the stages of the theatres and positioned wireless measuring devices at different locations in the seating areas. Participants were asked to adjust the volume of the sounds until they were just audible to them. Although impressive, the results weren't quite the stuff of legend. At Epidaurus, for instance, the sounds of a coin dropping or paper tearing carried throughout the theatre but weren't recognisable past the halfway point of the seats. A whisper could be heard only by those sitting in the front row, whereas words spoken at a normal volume couldn't be distinguished in the back rows of any of the theatres.[2]

In practice, the performers in Greek theatre were trained to speak boldly and with projection. They also wore masks, which may have further amplified their voices. Even people sat far back would have had no trouble making out the lines from their favourite plays by Aristophanes, Sophocles or Euripides. What's more, the sound quality of the amphitheatres in ancient times would have been far superior to what it is today. The stone surfaces of the theatres, made of polished marble, would have been smooth and shiny, not pitted and sometimes broken as they are now. Backdrops, too, almost certainly helped sounds from the stage be reflected out into the audience.

The great theatre of Epidaurus, designed by Polykleitos the Younger in the fourth century BCE.

The great theatre at Epidaurus, with a seating capacity of about fourteen thousand, has been especially admired for its remarkable acoustics, even now more than two millennia after it was built. The clarity with which voices and music can been heard, as far back as the fifty-fifth row, fifty metres above the level of the stage, is remarkable. It probably surprised even the Greeks who built it. They certainly made every effort to repeat the design in other places, albeit they never achieved quite the same level of perfection.

In 2007, researchers at the Georgia Institute of Technology set out to uncover the mysterious factor that makes Epidaurus so special. Others had suggested it might be the location, the degree of slope or possibly even the prevailing wind. As it turned out, it was none of these. The secret ingredient turned out to be the arrangement of the limestone seating in the theatre. The rows of seats form a corrugated surface, the shape and dimensions of which enable them to serve as an acoustic filter that transmits sound coming from the stage at the expense of surrounding acoustic noise. Lower tones, such as those from the murmuring of the crowd, are muted, whereas brighter tones from the performers are reflected back to audience members from the seats, all the way to the rear of the theatre, allowing them to hear clearly sounds coming from the stage.[3]

Another effect, of which Greeks at the time would have been unaware, was at work in theatres such as Epidaurus. Known as virtual pitch, it helps explain why, even far away, the full range of an actor's voice or a musical instrument can still he heard. Virtual pitch refers to the fact that, when we hear a combination of harmonics of a particular pitch and frequency, our brains insert the fundamental frequency into the sound we perceive.

Much of what the Greeks learned about acoustics came from trial and error. They discovered how sound behaves in amphitheatres by building these great structures, then tweaking them to improve the result for the audience. They learned about the behaviour of vibrating strings by plucking a monochord or a lyre. Aristotle, in the fourth century BCE, correctly proposed that sound travels by means of air in motion, though his theory, like many of those in Greek science, was based more on philosophy than experiment. Three hundred years later, the Roman architect and engineer Vitruvius hit upon the correct mechanism by which sound waves are transmitted and wrote extensively about the acoustic design of theatres. In the sixth century CE, Boethius put forward several ideas that related science to music, including the suggestion that our perception of pitch is related to the physical property of frequency.

For the birthplace and date of modern acoustics, though, we have to look to Renaissance Italy and the father-and-son pairing of Vincenzo and Galileo Galilei. Vincenzo we've already met in connection with the development of the equal temperament scale. In addition to being a talented lutist and well-regarded composer, he was also a skilled mathematician, who was the first to understand the relationship between the pitch of a note and the tension of the string on which it was played.[4]

Sometime before 1562, Vincenzo moved to Pisa where he married and fathered half a dozen children. One of the sons, Michelagnolo, grew up to become an accomplished lutist and composer in his own right. The oldest child, Galileo, we now know as one of the greatest astronomers and physicists who ever lived. Born in 1564, Galileo was encouraged by his father to experiment, make measurements and hold ideas up to scrutiny through systematic testing. In the basement of their house in Pisa, Vincenzo introduced the young Galileo to the science of sound and the power of experimentation. In place of philosophising and worldviews arrived at through thought alone, Vincenzo showed his son how to discover patterns in nature through careful observation and practical inquiry. Later in life, Galileo related to his biographer how the basement of his childhood home contained a forest of lute strings, each of different lengths and with various weights attached to them.

Better known for his experiments with pendulums and falling bodies, and for his discovery of the Moon's craters and the four largest satellites of Jupiter, Galileo also advanced our understanding of sound. Unlike his father, he devoted only a few pages to the physical study of acoustics in his *Discorsi*, published in 1638, but in that short space he established the link between pitch and what today we call frequency.

Pitch was something everyone had known about, going back into prehistory, because it's perceived directly by our ears and brains. Frequency, on the other hand, is a purely physical measure – the number of vibrations per second. There's no way we can count how many times, say a string, is vibrating back and forth when it's happening so fast, without some kind of instrument. Galileo didn't exactly have an instrument, but he came up with an ingenious method to make the frequency of sound visible. He scraped a chisel at different speeds across a brass plate, causing the chisel to make a series of regular grooves as it skipped along the metal surface. Galileo was able to observe how the pitch of the screech caused by the chisel varied directly with the spacing of the grooves. He didn't use the term 'frequency' but talked about the 'number of vibrations in the same time' and effectively gave the first experimental link between pitch and what would come to be known as the frequency of a sound.[5]

The pioneering work by Vincenzo and Galileo in musical acoustics was taken further by the French mathematician Marin Mersenne. Despite being a Catholic priest, Mersenne defended Galileo when he came under attack by the Church in 1633 for suggesting that Earth revolves around the Sun, rather than vice versa. The Frenchman repeated the experiments carried out by Galileo and his father but improved their accuracy. As a result, he was able to formulate, in precise mathematical terms, three laws governing the vibration of stretched strings. Mersenne's laws, which are still valid today, were published in his *Harmonicorum Libri* (1636) and formed the basis for modern musical acoustics. In the same book, Mersenne talks about the nature of sound, consonance and dissonance, modes of composition, the human voice, singing and all kinds of harmonic instruments – every aspect of acoustic and musical theories as they were then understood. But although this was a time when experimental science began to flourish, older notions didn't simply fade away. For Mersenne, the practical study of music and sound was still motivated by a Platonic belief in the harmonic basis of the universe.

With a firm foundation in place for the science of acoustics, other researchers began to build on that knowledge with the help of more advanced technology. Later in the seventeenth century, English physicist Robert Hooke became the first to produce a sound wave of known frequency, using a rotating cogwheel as a measuring device. Hooke began work on his wheel in March 1676, with help from the renowned clockmaker Thomas Tompion, following conversations with the music theorist William Holder. He had a long-standing interest in musical vibrations and a decade earlier had boasted to Samuel Pepys that he could tell how fast a fly's wings were beating from the sound they made. In July 1681, he demonstrated to the Royal Society his new device for producing distinct musical tones. It consisted of a brass wheel with numerous, regularly spaced teeth around its edge. As the wheel spun around, the teeth struck a card, making a noise: the faster the wheel turned, the more often the teeth hit the card and the higher the frequency of sound.

For the first time, there was a way to generate sound of a known frequency and to demonstrate empirically the correspondence between the human perception of pitch and the physical property of frequency. By fitting different wheels alongside one another on the same axis, Hooke was also able to verify frequency ratios for musical intervals, such as perfect fifths and fourths. And true to his word, he timed a fly's wing beats. By adjusting the speed of his wheel to produce a note that matched the humming sound of the fly, he arrived at his estimate. It turns out that a common housefly (*Mustica domestica*) flaps its wings about 200 times a second (200 Hz) – about the same pitch as a G or G sharp below middle C.

Hooke published his acoustic findings in 1705. Despite providing an objective measure of pitch, his wheel, as a practical tool for musicians, was quickly made redundant by the invention in 1711 of the tuning fork. More than a century went by before the French physicist Félix Savart took Hooke's invention a step further in his efforts to investigate the range of human hearing. By 1834, Savart was constructing brass wheels, eighty-two centimetres wide, containing as many as 720 teeth. By attaching a tachometer to the axis of the wheel, he could accurately measure the rate at which the teeth struck a card. His wheels, spun at high speed, could produce frequencies up to 24 kHz (24,000 Hz) – beyond the range of what any person can hear. It's been suggested, in fact, that Savart's wheel was the first artificial means of generating ultrasound. A unit of pitch interval has also been named after the Frenchman: 1,000 savarts is equivalent to three octaves plus a major third – about half the range of a modern piano. But its use has been almost universally superseded in music by that of the cent – one hundredth of a semitone in the equal temperament scale.[6]

Like many scientists who took an interest in musical acoustics, Savart enjoyed creating music as well. In 1819, he published the design for a violin in the shape of a trapezium. His aim was to improve the tone production of the instrument by giving its resonator a planar (flat) surface rather than a curved one. In this he was guided by the results of another pioneering acoustician, the German lawyer, musician and amateur scientist Ernst Chladni, based on his study of vibrating plates.

Chladni, who was born in the same year as Mozart (1756) and died in the same year as Beethoven (1827), investigated making musical tones visible in a solid material, having been inspired by earlier experiments carried out by Hooke. He spread fine sand (Hooke had used flour) over a metal plate and then drew a violin bow along one edge of the plate. The rosined hairs of the bow alternately gripped and slipped in rapid succession, making the plate vibrate and thereby causing waves to spread out across the plate until they were reflected from the other edges. The reflected waves became superimposed on new waves coming from the bow edge, resulting in symmetrical patterns of lines and curves where the plate remained stationary and the sand piled up. Many different 'Chladni figures' are possible depending on a variety of factors, such as the points of support and their location, the fraction of the distance along the edge where the bow makes contact, the frequency of the vibration (determined by the speed of the bow), and the shape and other properties of the plate itself.[7]

The patterns materialise when the plate vibrates at any of its special resonant frequencies. At these frequencies, the interacting waves on the surface form standing waves, with some locations where there's a lot of vibration and others where there's no vibration at all. Being a two-dimensional surface, with waves washing back and forth and interacting from different directions, the standing wave patterns on the plate can be very complex. But a vibrating string is simpler. As early as the late seventeenth century, French mathematician and physicist Joseph Sauveur noticed that certain points along a vibrating string remain still while others show a maximum movement. He called these points 'nodes' and 'loops', respectively, though today we refer to the loops as 'antinodes'.

Sauveur was an interesting character. As the scientist and essayist Bernard de Fontenelle pointed out: 'He had neither a voice nor hearing, yet he could think only of music. He was reduced to borrowing the voice and the ear of someone else and in return he gave hitherto unknown demonstrations to musicians.' Not only did he break new ground in acoustics, he invented the very word for this new branch of science: *acoustique*, from the Greek ακουστός, meaning 'able to be heard'.[8]

A better understanding of the science of musical acoustics inevitably had a powerful impact on all aspects of performance, from the construction of instruments (as we saw in Chapter 8) to the design of places where music was to be played. The acoustics of performance spaces has always been intimately connected with the kind of music that's played in them. Bach wrote his Brandenburg Concertos between 1718 and 1721 while in the patronage of Leopold of Cöthen, fully aware that they'd premiere in Leopold's court. So, he tailored the instrumentation for a small orchestra, capable of florid ornamentation, knowing that the work would be well suited to the small dimensions and low reverberation time of the room in which it would be performed. Composers have frequently adapted their work to the acoustics of where their work will be heard and, certainly in more recent times, concert halls have been built and equipped to provide an optimal listening experience for the audience.

In mediaeval times, churches were the sole venue for the highest forms of music in the West. But large, lofty spaces enclosed by bare stone walls have extraordinarily long reverberation times. Clap your hands inside a fourteenth-century cathedral and it will take several seconds for the sound to die away. Play fast, intricate music in such a place and the effort will be wasted, as the notes bounce off the hard surfaces, this way and that, overlapping and resulting in a cacophony. Even ordinary speech is hard to understand amid all the reverberation. Only one type of music really works in this kind of environment and, not surprisingly, it's the one that came to dominate the music of the Middle Ages: slow, monophonic chant. Through chant, the monastic choirs of that time could convey sacred texts in a way that took full advantage of the ethereal and spiritual effects produced by the resonance of these cavernous spaces.

Moving into the Renaissance and then the Baroque era and beyond, we find composers writing more specifically for different types of environment: livelier pieces involving fewer instruments for smaller rooms; grander works, with harmonies that change less frequently, where reverberation becomes more of a factor. When playing music indoors a little reverberation is good, but in some large Gothic cathedrals it can take up to nine seconds for the sounds from a pipe organ to die away, so several notes, though played separately, can end stacked on top of one another in a discordant pile-up. Modern concert halls are designed to avoid this problem, although performers recognise differences between reverberant spaces, describing some as being very 'live' and others as 'dead'. Optimum reverberation times depend on the type of music for which the hall is intended. Generally, good concert halls have a reverberation time between 1.8 and 2.2 seconds at mid-frequencies.[9, 10]

One of the features of the human ear is that it shows progressively greater discrimination against low frequencies as the sound gets softer. As a result, listeners far away from a source – say, an orchestra – may conclude that bass notes are too quiet in the overall mix. One way to offset this bias is to design auditoriums in such a way that the reverberation time for low frequencies is greater than for high frequencies. This gives a natural-sounding boost to bass sounds for audience members at the rear of the space. In practice, a longer reverberation time for low sounds often happens naturally, particularly if a lot of wood is used in the construction since wood absorbs higher frequencies more than lower ones.

The Royal Albert Hall in London is famed for its architectural grandeur and notorious for its dodgy acoustics.

Completed in 1871, this magnificent domed building has hosted everything from large classical concerts to modern pop and rock gigs. But it was designed by mid-Victorian royal engineers, not by architects and acousticians with an understanding of twenty-first-century science. It quickly became obvious that sounds ricocheted off its circular walls and, especially, its great dome, far too much. The Albert Hall was both overly reverberant and echoey.

Reverberation and echoes are related but aren't the same. Reverberation is the result of lots of reflected waves, which, when they're processed by the brain, are interpreted as a continuous, complicated sound. An echo, on the other hand, is a distinct but quieter copy of the original caused when a pulse of sound comes back to the listener after bouncing off a surface. If there's a delay of more than fifty milliseconds between the first and the second sounds reaching the ear, then they'll be perceived by the brain as separate rather than as one extended event. In terms of distance, an echo can be heard only if the reflecting surface is at least seventeen metres away.

From the very first concert following the opening ceremony, it was clear that the Albert Hall had a problem. Its concave, glass dome reflected and focused any sound that fell upon it, creating an obvious echo. A joke began to circulate that the venue was the only place where a composer could be sure of hearing their work played at least twice. In the century and a half since, numerous efforts have been made to improve the acoustic performance. Most recently, between 2017 and 2019, more than £2 million was spent on a major renovation, including a computer-designed sound system that can adjust the level of reverberation and echo perceived by different sections of the audience. Fifteen kilometres of cable were needed to connect the many new speakers installed![11]

Until quite recently, musical acoustics dealt only with instruments that make sounds by causing air to vibrate directly in some way – through a plucked string, for example. But through advances in technology, the twentieth century saw the rise of an entirely new form of musical sound production: one that depended on electricity and electronics.

CHAPTER 10

Experiments and Electronica

THROUGHOUT THE AGES people have experimented with new ways of making sounds and with new kinds of music that can be fashioned from those sounds. The harnessing of electricity fired the imagination of instrument makers, beginning in the late nineteenth century, and has led to the extraordinary proliferation of electronic devices today that allow any sound imaginable to be created from scratch.

If some accounts are to be believed, the first musical instrument to involve electricity appeared on the scene as early as 1748. It was the brainchild of Czech preacher and eccentric inventor Prokop Diviš, whose other scientific innovations included the first lightning rod (independently of Benjamin Franklin), various DIY weather machines, and methods of shock treatment for plants. Legend has it that to the field of music he contributed what he called the 'Denis d'or' or 'Golden Dionysus'. This clavichord-like instrument was reputed to have fourteen registers and a convoluted mechanism, housed in a wooden cabinet and operated by a keyboard, pedal and numerous stops. The iron strings – allegedly 790 of them – were struck, not plucked, and, through manipulation of the stops, could imitate the sounds of everything from a harp to an oboe.

According to the Belgian musicologist François-Joseph Fétis in his 1874 book *Biographie universelle des musiciens* ('Universal Biography of Musicians'), the Denis d'or was the first musical instrument to involve electricity, though, he adds, 'this was probably not an essential part of its action'. In fact, it seems far more likely, if the strings were electrified at all, that the intent was to spring a practical joke on the unsuspecting player rather than enhance the quality of sound. Fétis wrote: 'Diviš claimed that the performer could be shocked "as often as the inventor wished".'

No such mischief or controversy surrounds another eighteenth-century contraption that linked the musical with the electrical. In 1759, Belgian scientist, mathematician and Jesuit priest Jean-Baptiste Delaborde built his misleadingly named *clavecin électrique* ('electric harpsichord'). As Delaborde describes in a 1761 publication, the instrument, based on a contemporary warning-bell device, was essentially an electric carillon. A number of bells, two for each pitch, were hung from iron bars along with their clappers. The bars were charged from Leyden jars – an early form of capacitor. When the player pressed a key on the instrument, one of the bells of the corresponding pair was grounded, cutting it off from the charge source. The clapper then oscillated between the grounded and charged bells, ringing out the desired note. This 'marvel of the age' was admired by public and press alike, and not just for its musical tones. In the dark, its inventor wrote, 'the listener's eyes are agreeably surprised by brilliant sparks'.

The next century saw huge strides in our understanding of how electricity and magnetism worked. Many new devices took advantage of this enhanced scientific knowledge – among them, the phonograph (later known as the gramophone), invented by Thomas Edison in 1877. It provided the first means of playing back recorded sound and began a revolution in the music industry that continues to this day. Around the same time, the electrical engineer Elisha Gray came up with an early precursor of the modern synthesiser. The germ of the idea for his music-making device may literally have been child's play. Gray was watching his precocious nephew experiment one day with some of his uncle's equipment. In this era when health and safety were barely an afterthought, the boy had connected one end of a battery to himself and the other to a bathtub. Rubbing his hand on the bathtub's surface he managed to create a humming tone proportional to the electric current. Gray realised that, using the same principle, he could control sound from a self-vibrating electromagnetic circuit and thereby create a single-note oscillator. This formed the basis of his primitive sound synthesiser, which he called the 'musical telegraph'. He described how it worked in his patent notes of 1876:

> My invention primarily consists in a novel art of producing musical [...] sounds by means of a series of properly-tuned vibrating reeds [...] thrown into action by [...] keys opening or closing electric circuits. It also consists in a novel art of transmitting tunes so produced through an electric circuit and reproducing them at the receiving end of the line.

To show how the sounds could be transmitted, Gray built a simple receiver and loudspeaker called the 'washbasin receiver'. This was nothing more than a large telephone-like speaker made from an old metal washbasin mounted close to the poles of an electromagnet. By vibrating the washbasin, the receiver recreated and amplified the buzzing tone produced by the instrument.[1]

On an altogether more impressive scale was the telharmonium, patented in 1897 by Thaddeus Cahill, a lawyer and inventor from Washington, DC. Cahill had studied the physics of music at Oberlin Conservatory, Ohio, and had become fascinated by the possibilities of making music by electromechanical means. In 1902, he showed his first telharmonium to the famous Scottish physicist Lord Kelvin. At its heart were components called tone wheels – rotating disks, each of which produced a pure-tone sine wave. The disks were connected to a gearbox that was driven by an electric motor. Each disk had a certain number of smooth bumps at its rim, which generated a specific frequency as the disk rotated close to a pickup composed of a magnet and an electromagnetic coil. As each bump approached the pickup, it temporarily concentrated the magnetic field near it and thereby strengthened the field passing through the coil. This blip in field strength induced a current in the coil by electromagnetic induction. As the bump moved past, the concentrating effect fell again, the magnetic field weakened and an opposite current was induced in the coil. In this way, the frequency of the current in the coil depended on the rotational speed of the disk and the number of bumps.

Cahill's clever circuitry enabled the fluctuating current representing the note from one coil to be mixed with the currents from other coils representing other notes. The result was that he could combine a single fundamental frequency with one or more harmonics to produce a huge variety of note patterns. All of the musical activity was controlled by two performers sat at multiple sets of touch-sensitive, polyphonic keyboards, each with thirty-six notes per octave, tuneable to frequencies between 40 and 4,000 Hz. Cahill included additional tone wheels, the output of which could be used to colour the sound and produce a range of different instrumental timbres.

The telharmonium was well ahead of its time, and Cahill had high hopes for it. He envisioned its music being broadcast into hotels, restaurants, theatres and even private homes via the phone line. But it wasn't to be. The instrument was spectacularly large, weighing anywhere from seven to two hundred tons depending on its configuration, and eye-wateringly expensive with a price tag at the time of $200,000 – equivalent to more than $6 million today. In trials, phone users complained that strains of the telharmonium could be heard across the network, intruding on conversations. A few live performances of classics by composers such as Bach, Chopin and Rossini were given on it. But only three telharmoniums were ever built, none of which survive and, sadly, no recordings of this remarkable instrument exist.

Cahill's tone wheels were similar in concept to Savart's wheel, which we met in Chapter 9, but based on electromagnetic principles. Tone wheels would surface again, thirty years later, at the heart of one of the most iconic twentieth-century instruments – the Hammond organ – forming the basis for its unique sound.

In 1920, another revolutionary sound-making device emerged that looked more like a school science project than anything that could be used to create music. Invented by, and eventually named after, Russian cellist and physicist Leon Theremin, it consisted of a cabinet, containing a loudspeaker, from which protruded two antennae – a horizontal loop to the left and a vertical pole to the right. Movements of the player's left hand around the loop altered the volume of the sound, while the right hand controlled pitch, according to the distance from the pole. The theremin exploits the capacitance properties of the human body and works in the same way that someone moving near a radio can alter the reception. It was the first example of a purely electronic instrument and the first designed to be played with no physical contact.[2]

Alexandra Stepanoff playing the theremin on NBC Radio, 1930

The theremin emits a continuous monophonic tone, similar in character to a violin but with an ethereal, otherworldly character. Almost inevitably, it became a staple for conjuring up alien sound effects in classic science fiction films of the 1950s, such as *It Came From Outer Space* and *The Day the Earth Stood Still*. Similar sounds come from the electro-theremin or tannerin, developed by trombonist Paul Tanner and amateur inventor Bob Whitsell in the late 1950s to mimic the theremin. In this case, the player controls the frequency by moving a slider that's attached to a pitch knob, which in turn is connected to a sine-wave generator. The eerie effect on the Beach Boys' 'Good Vibrations', often assumed to be produced by a theremin is, in fact, made by a tannerin.Also taking inspiration from the theremin was an instrument developed by French cellist and World War I telegrapher Maurice Martenot. In the early 1920s, Martenot and Theremin met and discussed possible improvements to the Russian's eponymous device. The result was the ondes Martenot ('Martenot waves'), first heard in April 1928 playing its inventor's *Poème Symphonique*. Its haunting tones are controlled by means of a finger ring attached to a length of ribbon, which is pulled up and down a keyboard to determine pitch while the left hand operates a volume control. Like its cousins, the theremin and tannerin, it has at its heart a variable oscillator – a sine-wave generator – that can produce pure tones and a distinctive glissando (a glide from one pitch to another). But different playing techniques used with the ribbon and keyboard give it a greater range of timbre.

The ondes Martenot has appeared in scores by Boulez, Honegger, Ibert, Milhaud, Varese and others. Six of them were needed by Olivier Messiaen in his *Fêtes des belles eaux* ('Festival of the Beautiful Waters'), written for the 1937 Exposition Internationale des Arts et Techniques dans la

Vie Moderne, a light-and-water show on the Seine, in Paris. Like the theremin, it can be heard on the soundtrack of many horror and science fiction films, including *Mars Attacks!* and *Ghostbusters*. Radiohead's Jonny Greenwood used it on the band's *Kid A*, *Amnesiac* and *Hail to the Thief* albums, and he also wrote the piece 'Smear' for two ondes Martenot and the London Sinfonietta.[3]

While some composer–inventors experimented with the new world of musical electronics, others explored scales and systems of tuning that went far beyond the Western norm. Beginning in the 1930s, American composer Harry Partch built unique instruments, striking in appearance, to play pieces that made extensive use of microtones (the subject of Chapter 12). From wood, glass, metal and bamboo he sculpted such exotica as the chromelodeon, the adapted viola and the Ptolemy (a bellows-reed organ) that could play 29-, 37- or 43-tone octaves. His extraordinary Cloud-chamber Bowls consist of large glass gongs of various sizes suspended in a wooden frame and played with mallets. Partch initially created the Bowls in 1950 using Pyrex carboys discarded by the Radiation Laboratory at the University of California, Berkeley.[4]

The 1930s also saw the rise of some of the most iconic instruments of popular music in the twentieth century. The first true electric guitar was put together by American inventor and vaudeville musician George Beauchamp, with the help of machinist Adolph Rickenbacker, for the Electro String Instrument Corporation. Nicknamed the 'frying pan', it had a flat, circular body connected to a long neck and was a lap steel guitar designed to take advantage of the popularity of Hawaiian music at the time. Although Beauchamp and Rickenbacker began selling the Frying Pan in 1932, its design wasn't patented for another five years, which allowed other companies to produce guitars using the same technology.[5]

At the heart of the electric guitar is the pickup. Magnets in the pickup convert the vibrational energy of the steel strings into tiny pulses of alternating current. This current is then boosted by an amplifier and fed to a loudspeaker, allowing the normally quiet guitar to be heard above other, noisier instruments. Electrification meant that the guitar could hold its own even when playing with the loud brass section of big bands of the swing era. Later, in the 1940s and 1950s, when bluesmen like Muddy Waters and Howlin' Wolf moved north from their homeland in the American South, amplification enabled them to compete with raucous crowds in the bars and clubs of Chicago.

Some famous names became established early in the rapidly expanding world of electric guitars. In 1934, Electro String Instrument was rebranded as Rickenbacker. Two years later, legendary acoustic manufacturer Gibson joined the field with its Electric Spanish Guitar series 150, more commonly known as the ES150 – the number representing its $150 price tag at the time. The ES150 was a hollow-bodied electric with pickups in place of the traditional sound hole and f-holes on the guitar's sides. These early electrics had one major drawback: they suffered from feedback when played at high volume, partly due to reverberation within the hollow space. In an effort to mitigate the problem, guitar manufacturers sometimes opted to make the body's wood thicker, leading to a semi-hollow body design with a solid piece of wood running through the middle of the instrument.

American jazz and country guitarist Les Paul began work on what would become the first solid-body guitar in 1939. After Paul befriended Epaminondas 'Epi' Stathopoulo, owner of Epiphone music company, Epi gave him keys to the firm's New York factory so that Paul could work on his design after hours. The result was a bizarre-looking creation, dubbed 'the Log', which consisted of a four-by-four-inch length of pine to which was attached a neck, bridge and pickups. It sounded fine but caused bewilderment among audiences whenever they saw Paul playing it live. In response, Paul asked a luthier to slice an Epiphone hollow-body guitar in two and attach the halves to either side of the solid slab to give the Log a more conventional look.

After World War II, the first of the modern solid-body guitars began to appear. American inventor and designer Paul Bigsby brought in several innovations that became standard fixtures on some future makes of guitar. His 'vibrato tailpiece' – also known simply as a 'Bigsby' – was a mechanical device that allowed musicians to bend the pitch of single notes or entire chords with their pick hand. In 1948, Bigsby created a guitar for the country and western star Merle Travis that featured a headstock with six tuners in a row – a design that would quickly find its way into the iconic electric guitars of Leo Fender, including the Telecaster and the Stratocaster.

Unlike Les Paul, Leo Fender wasn't an accomplished guitarist. He played saxophone and piano, but his real passion was electronics. As a teenager, he visited his uncle's automotive–electric shop in Santa Maria, California, and was immediately intrigued by a radio his uncle had built from spare parts. It helped spark a lifelong interest in both electric instruments and amplifiers. During the war years,

Fender – ineligible to serve, having lost the sight in one eye as a child – teamed up with 'Doc' Kauffman, an inventor and lap steel player. Together they designed and built electric Hawaiian guitars and amplifiers before Doc pulled out of the business and Fender formed his own company.

The late 1940s saw an increase in popularity of roadhouses and dance halls, and a growing demand for louder, less expensive and more durable instruments. Musicians also wanted 'faster' necks and better intonation to perform what country players called 'take-off lead guitar'. Fender was quick to grasp the potential for an electric guitar that was easy to hold, tune and play, and wouldn't feed back at dance hall volumes. In 1948, he finished the prototype of a thin solid-body electric. A single pickup version of this went on sale in 1950 as the Fender Esquire, followed by a two-pickup model, initially called the Broadcaster but renamed the Telecaster a year later.

The Telecaster proved so successful that it encouraged Gibson to enter the solid-body marketplace too. Les Paul was enlisted to help spearhead the design of their new product and, in 1952, the Gibson Les Paul was released. Along with Fender's Stratocaster it remains one of the most instantly recognisable instruments in the hands of rock guitarists the world over.[6]

Just as electric guitars first began to appear in the 1930s, so too did the first commercially available electric organs. In 1933, Laurens Hammond in the United States unveiled such an organ, the distinctive sounds of which continue to be heard in pop and rock music today. Hammond was a gifted inventor who graduated with an honours degree in mechanical engineering from Cornell University in 1916. After World War I, his invention of a silent, spring-driven clock gave him enough capital to strike out on his own, and in 1928 he founded the Hammond Clock Company, which made a range of electric clocks, driven by another of his inventions – the synchronous electric motor. During the Great Depression he saw music as a field in which to diversify and market new products. In developing the electric organ, Hammond turned for inspiration to the underlying principles of Cahill's ill-fated telharmonium. Aided by his company treasurer (and church organist) William Lahey, Hammond used his engineering skills to develop an electromechanical system of tone-wheel generators coupled to a keyboard. The Hammond tone-wheel organ was patented in 1934 and the Model A went into production in 1935, with Henry Ford and George Gershwin among the first customers.

The now-legendary B3 model came out in October 1955 and quickly became a favourite with musicians of all genres for its distinctive sound and versatility. The B3 is housed in a large wooden cabinet on four spindle legs, with separate power amplification and speaker systems. It's equipped with a pair of 61-note keyboards and a 25-note radial removable pedal board. The sound generated by the instrument is controlled by a series of rocker switches and drawbars. These drawbars lie at the heart of the Hammond sound, allowing the player to build up rich timbres by combining pure tones in different ways – just as a church organist uses stops to combine pipes of different lengths.

Another defining feature of the Hammond sound was invented not by Hammond himself but by Don Leslie. The Leslie speaker was developed to overcome the shortcomings that Leslie felt to be inherent in the sound of the Hammond organ. Noting that the instrument sounded much more impressive in large halls, Leslie began to experiment with ways of injecting reverberation and motion into the sound. The result was the rotating speaker system that bears his name.

The Leslie speaker cabinet looks like a large chest of drawers and contains a forty-watt tube amplifier, a rotating treble horn and a rotating bass speaker. It was designed to produce two particular effects – the 'chorale' (when the speakers rotate slowly) and the 'tremolo' (when they spin rapidly). However, a third effect, exploited by many players, was made possible by disconnecting the slow motors in order to make the change from 'chorale' to 'tremolo' much more exaggerated.

Hammond went on to produce different models of electric organs, all of them based on the original tone-wheel technology, until 1975. The company continued to make organs using electronics and digital technology, but it went out of business in 1986, only to be bought by the Suzuki group the following year. New instruments, such as the B3 and the XK3, use modern digital technology to emulate the vintage sound so beloved by many keyboard players. The characteristic Hammond and Leslie combination can be heard on countless jazz, rock, gospel and blues recordings, including such classics as Procul Harum's 'A Whiter Shade of Pale'.[7]

Far removed from any mainstream genres of music, even those energised and amplified by the new electrified instruments, was the upstart form known as *musique concrète*. This began in Paris in 1948 through the efforts of radio engineer Pierre Schaeffer and composer Pierre Henry, in the Studio d'Essai at Radiodiffusion Studio d'Essai Française (RDF). It didn't use any instruments, traditional or electronic, but instead involved manipulating real-world ('concrete') sounds that had been recorded on tape. Schaeffer would speed up or slow down recordings, reverse some sections and repeat others. He discovered that sounds took on a different character when the initial attack, or rise, of the sound was taken out. His first piece, *Étude aux chemis de fer* ('Study with Trains'), was based on the recorded sounds of train engines, wheels and whistles, and broadcast on RDF with an introduction by Schaeffer, who called it a *concert de bruits* ('concert of noises'). Schaeffer and Henry soon became musical collaborators and produced a number of *concrète* works, including *Symphonie pour un homme seul* ('Symphony for a Lone Man'). This piece made use of human sounds, such as breathing and vocal noises, as well as the sounds of percussion instruments, doors slamming, piano and orchestral textures. In the 1950s, *concrète* composers perfected methods, such as cutting, splicing and looping, which soon became familiar throughout the industry, for manipulating sounds recorded on magnetic tape.[8]

Others began fusing the ideals of pure *concrète* works with electronically generated music. In 1958, French composer Edgard Varèse combined the found sounds of *concrète* with synthesised electronic sounds to create his *Poème èlectronique*, debuted at the Belgian World's Fair through a system of 400 loudspeakers. Meanwhile, German composer Karlheinz Stockhausen, who briefly worked in Schaeffer's studio in 1952, pioneered *electronische Musik*, emphasising electronically generated and modified sounds rather than tape manipulation. These various experimental forms eventually found their way into pop and rock music of the late 1960s and beyond. Sound recorded from non-musical sources can be heard, for example, in tracks by the Beatles and Pink Floyd, whereas synthesised and 'sampled' music became increasingly common in everything from progressive rock to disco and hip-hop.

Also instantly recognisable on some tracks, especially in early progressive and psychedelic rock recordings from the mid-sixties to mid-seventies, was the Mellotron. This had its roots in an earlier electromechanical instrument, first manufactured by Harry Chamberlin in California in about 1952 and named after him. The Chamberlin consisted of a small keyboard, spanning about three octaves, which was used to activate loops of AMPEX magnetic tape on which were prerecorded the sounds of string, woodwind or brass instruments. Each loop ran for about seven or eight seconds before a spring mechanism returned the length to its beginning. Different sounds or voices were selected by using controls next to the keyboard to shift the position of the tape heads.

In 1962, Chamberlin's agent, Bill Fransen, visited Bradmatic, a small company in Birmingham, England, run by brothers Bill and Les Bradley, which made Spitfire parts and other items, including playback heads, which Fransen was interested in sourcing. The Bradleys realised they could improve on the Chamberlin and turn it into a commercially viable product – the Mellotron. Their Mark I appeared in 1963 but, although it was technically superior to the Chamberlin, it was also temperamental. Design improvements led to the Mark II, produced from 1964 to 1967. During the early part of that period, Mike Pinder worked at the Bradley's factory before going on to become a founding member of the Moody Blues. Pinder's skill at playing the instrument – not to mention his technical wizardry at repairing it when it broke down – helped the Mellotron feature so distinctively in the Moodies' seminal albums. King Crimson and Genesis were among other bands of the time who added the Mellotron's orchestral strains to the more familiar rock instrumentation of guitars and drums to add depth and power to their sound.[9]

In retrospect, the Mellotron can be seen as the earliest example of a sampler – a device that stores previously existing sounds and plays them back when needed. But whereas the Mellotron is a strictly analogue instrument, samplers today are based on digital technology and play a central role in music production. In digital sampling, a recorded analogue signal is analysed and transformed through an analogue-to-digital converter into digital information that can be easily stored, manipulated and replayed. On playback, the process is reversed and the digital data are passed through a digital-to-analogue converter to recreate the original waveform.

The digital revolution in music can trace its roots to the late 1950s. At Bell Labs, in 1957, electrical engineer Max Mathews wrote MUSIC, a computer program for generating digital audio waveforms through direct synthesis. It was among the first programs that could make musical sounds on a digital computer and the first program to be widely accepted in the music research community as being useful for this task. In 1961, Mathews arranged a rendition of the song 'Daisy Bell: Bicycle Built for Two' with a computer-synthesised human voice, using technology developed by several of his colleagues. Author Arthur C. Clarke happened to be visiting a friend, John Pierce, at the Bell Labs Murray Hill facility at the time of this remarkable speech-synthesis demonstration. He was so impressed that he later asked Stanley Kubrick to use a recording of it in *2001: A Space Odyssey*, in the scene where the HAL 9000 computer sings while its cognitive functions are progressively disabled.[10]

Pierce, who'd played a leading role in the development of the Telstar 1 communications satellite, and Mathews worked together, running their music software on Bell's most powerful number cruncher – the IBM 7090. Equipped with a memory of just 32,000 36-bit words, it was hardly in HAL's class, yet the music it generated, played through a digital-to-analogue transducer, was revolutionary for its time. It led to the first digital recording, an album called *Music from Mathematics*, issued by the Decca Record Company in two parts in 1960 and 1962. According to a review in the January 1963 issue of the journal *Science*: '*Music from Mathematics* is an exciting record that presages an area of endeavour likely to expand greatly in coming years… To those who are curious about things new in music and to those who wish to see some of the paths music may take, this record is highly recommended.' It adds, 'It is not surprising to learn that musicians compose more convincing music than scientists, but it is also certain that scientists can contribute greatly to music through the creation of new knowledge and tools for musicians.'[11]

Later in the sixties came another music breakthrough courtesy of the revolution in digital and microchip technology – the synthesiser. Spearheading this development was American engineer Robert Moog (rhymes with 'vogue') who'd previously built and sold theremins while working towards his doctorate at Cornell University. In 1964, Moog met Herbert Deutsch, a composer and music teacher at Hofstra University, from whom he learned of the rising demand for more practical and affordable electronic-music equipment. In response, he devised the voltage-controlled oscillator, an electronic circuit that uses voltage to control pitch. It's the central component of the Moog synthesiser, which first went on sale in 1965. In its original form it consisted of a keyboard and accompanying pitch wheel, attached to a large cabinet with dauntingly complex-looking panels of knobs, wheels and levers arranged into 'modules'. When wired together in various ways, these modules could produce a fantastic range of previously unheard sounds.[12]

The scope and potential of the Moog Modular Synthesiser, as the first commercial model was known, were spectacularly demonstrated on the 1968 Grammy award-winning album *Switched-on Bach*. Containing a collection of pieces by Johann Sebastian Bach, arranged by American composer Wendy Carlos and played on the Moog by Carlos and Benjamin Folkman, *Switched-on Bach* reached number ten on the Billboard 200 chart in the United States and topped the classical album charts from 1969 to 1972. It played a key role in bringing synthesisers into popular music, whereas before they'd been seen largely as experimental instruments. Among the first to take advantage of the Moog for pop were, surprisingly, the Monkees, who featured it on their 1967 album *Pisces, Aquarius, Capricorn & Jones Ltd*. The Beatles, on their *Abbey Road* album, alongside such diverse artists as Bob Marley, Chick Corea, Rush and Yes, were quick to incorporate synthesisers into their work. Perhaps most famously, Keith Emerson, of Emerson, Lake & Palmer, brought an enormous Moog synthesiser on stage and often played it at the same time as a Hammond organ – one hand on each keyboard. In Germany, the synthesiser spawned a whole subgenre of progressive rock – krautrock – represented most prominently by Tangerine Dream, Can and Kraftwerk.

Digital sampling was added to the armoury of electronic musicians. Once a sound has been sampled – recorded and digitally stored – it can be loaded into a synthesiser's bank of sounds and played or manipulated at will through the press of a key. As the electronic-music industry rapidly grew and evolved, it became clear that some standard was needed so that the various products coming on the market could effectively communicate with one another. Major companies, including Roland, Yamaha, Korg, Kawaii and Sequential Circuits, collaborated and in 1983 announced the Musical Instrument Digital Interface (MIDI) as a standardised means of synchronising electronic instruments.

The 1980s also saw the emergence of microcomputers powerful enough to handle digital audio editing. The software needed to integrate the functions of a studio on a single computer followed soon after, leading to the digital audio workstation (DAW). Many major recording studios finally went digital after Digidesign introduced its Pro Tools software in 1991, modelled after the traditional method of handling signal flow in analogue recording devices. Today, anyone with a laptop and DAW software, some of which is free, can effectively set up a studio at home to record and create their own original music.

Electronics has made music a democratic art. Beginning in the late 1950s, amplification – especially of electric guitars – gave young people with a message, but not necessarily any musical training, a way to project themselves and their sound. Electronic keyboards like the Hammond, Mellotron and synthesiser put powerful new effects at the disposal of individuals and small bands: no longer did you need an orchestra or radiophonic workshop to sound like one. With the coming of affordable yet astonishingly capable software and high-performance computers, linked to MIDI devices and samplers, it's now possible to make great music in a backroom at home. Thanks to the electronics revolution, anyone, anywhere, given some basic equipment and YouTube lessons, has the potential to unleash their musical imagination and, through the Internet, show their work to an audience worldwide.

CHAPTER 11

Did the Beatles Play Out of Tune?

LET'S SAY YOU want to play along with an old Beatles song – maybe 'Ticket To Ride', which was issued as a single in 1965. You've got your instrument ready, whether it be a guitar or a piano. If it's a guitar, you've made sure it's properly tuned to concert pitch (A4 = 440 Hz) using an electronic tuner. You know what chords are used in the song, and you start playing along with this old recording. Very quickly it's clear something isn't right. You're playing exactly the same notes as the Beatles but it sounds awful. What on Earth is going on?

It turns out that a lot of early Beatles songs were recorded at or about A4 = 435 Hz, so if you're tuned to concert pitch, A = 440Hz, you're going to be a bit sharp – playing at a slightly higher frequency than on the recording. If you hear pure tones of 440 and 435 Hz separately, in isolation, it's hard to tell any difference between them. But played together they sound bad because you can hear beats – the result of two waves of slightly different frequency interfering with each other. The number of beats per second is the difference in frequencies between the two pitches.

To be in tune with the Beatles on most of their early stuff you have to 'detune' to around 435 Hz or, in some cases, as low as 432 Hz – in any case, several cycles per second lower than standard concert pitch. Now, if you're playing on your own or with a band doing a cover, it doesn't matter about making this adjustment as long as your instruments are in tune with *each other*. But Beatles tribute bands that want to sound as close to the original as possible may go to the trouble of tuning down to get a near perfect match with what's on the record.

Television performance by the Beatles in Treslong, Hillegom, the Netherlands, on June 5, 1964.

Why wasn't the early Beatles' stuff played at 440 Hz? The fact is, a lot of early pop and rock recordings aren't at concert pitch. Sometimes that's because the band tuned to whatever fixed point of reference happened to be in the studio – usually the upright piano that was provided for anyone doing a recording. So, if the studio piano wasn't tuned to 440 Hz, neither would be the band. Some of the songs on the Rolling Stones' *Exile On Main Street* album aren't at 440 Hz for that very reason, and it may also have been the case with the Beatles' early recordings. But it's complicated. On the *Help* album, for instance, released in 1965, most of the songs are at, or very close to, 440 Hz. However, 'Ticket To Ride' and 'You've Got To Hide Your Love Away' are somewhere between a half step (a semitone) and a quarter step below standard pitch.

The Beatles also *deliberately* issued songs with different tunings, especially in their later albums such as *Revolver* and *Sergeant Pepper's Lonely Hearts Club Band*. In those days, everything was recorded on tape, which could be sped up or slowed down in a technique called vari-speeding. The Beatles would record some instruments and vocals a semitone or two away from the final mix tonality in order to alter the timbre or quality of the sounds that appeared on the record. They'd even sometimes go as far as recording at half speed. The solo piano part on 'In My Life', which is actually played by their producer George Martin, was intentionally recorded at a much lower speed and then sped up so that Martin could avoid making mistakes. Of course, when the Beatles and other bands started playing along with orchestral instruments, as they eventually did, then they had to make sure that they were tuned exactly to the conventional 440 Hz.

Sometimes musicians deliberately tune their instruments so as not to be in equal temperament, that is, they'll purposely shift the pitch of one or more notes to achieve a certain effect. John Lennon said that when tuning his guitar he'd always tune the high E string slightly flat because when playing open chords, he believed, it gave the higher notes a softer sound.

The term 'sweetening' is sometimes used to describe this technique of altering the relative tuning of certain notes of the scale. Like all forms of tuning, it involves a compromise: the sound of some chords, or note combinations, may be improved but at the expense of others.

Singer-songwriter James Taylor has, over time, come up with a very specific form of sweetened tuning that involves slightly retuning every one of the six strings on his acoustic guitar away from standard A4 = 440 Hz. As he points out, the guitar isn't a precise instrument. Variations occur as you play up and down the neck or if a capo is used as a way of transposing the key. Also, different strings have different tendencies in terms of how they harmonise. Over time, Taylor has devised a system of tuning the guitar to address some of these vagaries and conflicts that are part of the instrument. Bass strings, for example, especially when played hard, tend to ring sharp, so Taylor tunes them slightly flat, relatively speaking. The B string, too, he believes needs to be tuned flatter relative to its neighbours. Using an electronic tuner that can measure in cents – one hundredths of a semitone – he recommends detuning all the strings by amounts ranging from minus three cents for the high E string to minus twelve cents for the low E string, these being his personal preferences.[1]

No such attention to fine detail was involved in the production of Led Zeppelin's 'Heartbreaker' on their 1969 album *Led Zeppelin II*. The song opens in standard A440 tuning, but Jimmy Page's legendary solo, which starts about two minutes in, is sharp of this. As Page has pointed out, his solo was recorded at a different time and in a different studio, but whether the discrepancy was due to the tuning of his guitar or the equipment in the studio we'll probably never know.

Intentional adjustments to pitch can be made at either the production or the performance stage. Through years of experience of producing and audio engineering, George Martin found that instruments that are perfectly in tune with each other make for a rather flat-sounding recording. When John Lennon and George Harrison were both playing their guitars on a song, Martin would purposely shift one of the guitars slightly out of tune with the other to give a more expansive sound in what, in those days, was a mono recording. The same principle is behind double-tracking vocals and so-called stretched tuning of pianos and other wire-strung instruments.

Today, as we've seen, almost all Western music uses 12-TET tuning. This means that every interval in the 12-TET scale involves the same jump in frequency ratio – equal to the twelfth root of 2, or 1.05946… A sharp is 1.05946 times the frequency of A, B is 1.05946 times the frequency of A sharp and so on. Only notes that are an octave apart are exactly consonant in terms of their frequency. All other intervals are dissonant to some extent – either a bit sharp or a bit flat – but not so much that it bothers us, because we've grown accustomed to the aural compromises that 12-TET involves. Because the gap between every note in 12-TET is the same, from an objective standpoint you wouldn't expect there to be any real difference in how each key feels. In other words, a piece that's in the key of E minor, for example, should, in theory, sound no different in terms of mood or emotional impact than the same piece in D minor. The gaps between all the notes, measured in frequency ratios, would be the same.

This wasn't the case, however, before equal temperament came along. In other systems of tuning, the sequence of gaps between notes differ depending on the key. This led to certain characteristics becoming associated with the various keys, A to G, and whether major or minor. C major, for instance, was thought of as being completely pure, simple and naive. D major had a reputation for being the key of triumph and rejoicing, and therefore became a favourite for symphonies and marches. E flat major was associated with love and devotion, F minor with depression and, according to one musical text, 'groans of misery and longing for the grave'.

Today, with 12-TET, there's no distinction between the jump in frequency ratio of, say, a major third in one key and a major third in another. But in earlier times, the major third in one key might be close to a perfect fourth in another or a minor third in yet another. A change in key, therefore, could result in a very noticeable shift in the mood of the piece. Although to a large extent these variations in emotional feel of different keys have been eliminated by equal temperament, this doesn't mean that a musical composition will sound the same, or as pleasing to the ear, in every key.

Most obviously, a singer-songwriter who intends to perform a piece they've written themselves will choose a key and, with it, a span of notes which suit their voice type and range. The same is true of any choral work: the key needs to be carefully chosen so that all the notes in the composition are comfortable to perform. Composers must also bear in mind the qualities of any instruments their works call upon, recognising that the characteristic sound a cello or a clarinet, for example, makes will differ depending on the register in which it's being played. The choice of home key, and any modulations between keys, is strongly influenced by the instruments (including the human voice) for which a piece is intended.

Another factor in the choice of key is that music – especially a vocal in pop and rock music – sounds brighter and more exciting if it's pitched higher. Transposing down, on the other hand, enhances the sonic darkness. But when dealing with low frequencies, music producers have to be careful because in the bass region of the spectrum there's a tendency for notes that are close together to start interacting. This can give rise to audible beats as the frequencies clash.

When it comes to notes played on real musical instruments, rather than pure tones that sound at a single frequency, it's complicated. Press a key on a piano and, in addition to the fundamental frequency, a whole series of harmonics or overtones will be produced. This complexity of sounds can affect our perception of the pitch of notes. For example, a C played towards the higher end of a piano may sound slightly flat compared with a C a couple of octaves lower. So, whereas the definition of an octave is based on the assumption of a fixed linear relationship – namely, that jumping up an octave doubles the frequency – this doesn't necessarily square with our actual pitch perception, which is non-linear. This is why in so-called stretched tuning, two notes separated by an octave, whose fundamental frequencies theoretically have an exact 2:1 ratio, are tuned slightly farther apart.

Musicians and composers continue to debate the relevance today of 'key psychology' or 'affective musical key characteristics' – the notion that different keys conjure up different moods, which we discussed in Chapter 4. But it's clear that the main determining factor in how a song or instrumental piece influences the listener isn't so much the key as the composition itself – its tempo, melody, harmonies and orchestration. Emotions are evoked not because the key is D major, E major or any other, but primarily because of the contrast between tension and release.

One way to create tension is through dissonance. This doesn't mean playing out of tune – although that certainly would be dissonant – but rather juxtaposing notes in the scale that very obviously don't harmonise. For example, two notes separated by a minor second – one semitone – clash if played together, yet if played one after the other, they can be used to create a sense of anticipation or even dread. The most famous example of this is in the theme tune to the movie *Jaws*. The opening notes alternate between E and F, going up and down a minor second, faster and faster as the shark closes in. Just two notes repeated, over and over, with *accelerando* (increasing tempo), is all it takes to build the feeling of unease to an almost unbearable level.

Tension of a different kind arises from the first two notes of the song 'Maria' from *West Side Story*. Starting at the root note 'Ma' we jump up an interval of three whole tones, otherwise known as a diminished fifth, an augmented fourth or a tritone, to 'ri'. For that moment, at the very beginning of the song, we're left hanging briefly in a state of suspense before the next note comes to the rescue, shifting up a further semitone – a perfect fifth from the starting note – and thereby returning us to stability with the final syllable 'a'. (Incidentally, exactly the same three notes are played at the start of *The Simpsons* theme tune!) The tritone is notorious throughout Western music for its ability to unsettle the listener, and Leonard Bernstein, well aware of this, uses it liberally throughout the score for *West Side Story*. Whenever he wants to hint at violence or approaching danger, he leaves the tritone unresolved, and where he wants to suggest optimism he follows the tritone with a resolved note or chord. At the very end of the musical, he leaves the audience in suspense, musically and emotionally, by having two alternating tritones play out against each other.[2]

Often in life we look forward to times when we can just sit back and relax. But too long spent with our feet up, when nothing much happens, gets to be dull after a while. Temporary stress and unease can be exciting, especially if it's followed by a period of calm. It's why we want novels to have ups and downs, tension and resolution. It's why we go on rollercoasters: it's a thrill. Feeling unsettled for a while takes us out of our comfort zone, and then we get to ease back at the end. Dissonance is an important tool of the composer for the simple reason that tension, as in life itself, makes music interesting.

Of course, that's not how the Church saw it back in the day. The last thing that sacred music was supposed to be was jarring on the ear because that could be construed as being offensive to God. The chants sung throughout the Middle Ages were intentionally and uniformly constructed to be austere, pure and, above all, consonant in order to glorify the Almighty. Perfect fourths and fifths were the order of the day, and as for any dissonant interval, perish the thought. That's not to say, though, that tritones were unheard of in mediaeval times. Pérotin, that giant of the Notre Dame school of polyphony in the thirteenth century, lightly dusted tritones throughout his *Viderunt Omnes*, an organum for four voices based on Psalm 98.

A myth has grown up that terrible things could happen to anyone who defied the Church by blatantly inserting dissonances into a sacred composition – and there's nothing more dissonant than a tritone. But no records exist of a composer from mediaeval or Renaissance times who'd slipped an occasional tritone into one of their pieces actually being burned at the stake or even sternly admonished. While the ecclesiastical powers that be were generally of the opinion that consonances were divine and dissonances just the opposite, intervals such as the tritone weren't actually banned or deemed heretical in the Middle Ages. It's just that until greater musical sophistication came along with the arrival of polyphony, no one thought to use dissonance constructively.

The term *diabolus in musica* – 'the devil in music' – was coined to describe the tritone. But it wasn't the Church that came up with this phrase as a way of implying a dark, satanic sound. One of the earliest references to it is by the eighteenth-century Austrian music theorist Johann Joseph Fux in his celebrated book on counterpoint *Gradus Ad Parnassum* ('Steps to Parnassus'). And it's generally recognised that Fux, far from making a religious comment, used the expression as a way of recommending dissonance as a tool for adding colour and interest to music.

Even during the Baroque period, composers were routinely adding a little sonic spice in the form of dissonant intervals. By the middle of the nineteenth century, some composers were going out of their way to exploit the tritone for creating tension or, in some cases, overtly implying dark forces and events. In 1849, Franz Liszt wrote his *Dante Sonata*, a piece for solo piano based on Dante's epic drama of heaven and hell, *The Divine Comedy*. Portions of all three books in the 14,233-line poem are represented in the single-movement sonata, including the famous *Inferno*. For the descent into the fires of hell, Liszt, in the opening bass line, draws heavily on the tritone, leveraging its traditional association with the devil.[3] It's also the interval of choice at the start of the violin section of Camille Saint-Saens's *Danse Macabre*. The work is based on a poem by Henri Cazalis that describes Death himself arriving at a graveyard at midnight on Halloween and summoning the skeletons to dance with him and his sinister, tritone-laced fiddle tune. Saint-Saens dials up the dissonance even further by calling for the violin's E string to be detuned to an E-flat – a technique known as *scordatura*. This turns the usually sonorous perfect fifth on the violin's open strings of A and E into a discordant tritone of A and E flat. With one string out of tune, it also has the effect of making the work fiendishly difficult to play.

In the hands of Richard Wagner dissonance became a compositional form in its own right. A heavily dissonant chord erupts in only the second bar of his opera *Tristan und Isolde*, immediately whipping up tension in the listener. So unusual is the chord when contrasted with the implied key against which it's played that it's become known as the 'Tristan chord'. It's often taken to be a watershed in music – the place from which twentieth-century atonality first sprang. Having announced his intention to break with standard musical practice, Wagner goes on to treat his audience to almost four hours of relentless dissonant opera before finally drawing to a consonant resolution in the final few bars. In a section of his *Götterdämmerung* – the fourth and last part of his Ring cycle – he makes full use of the tritone to portray a pagan scene.

In 1913, Igor Stravinsky wrote his groundbreaking piece *The Rite of Spring*. Taking his lead from Wagner, Arnold Schoenberg and Anton Webern, Stravinsky's ballet and orchestral work proved so divisive in its use of dissonance that it caused a sensation upon its premiere. Labelled 'a laborious and puerile barbarity' by one newspaper, it's now recognised to be one of the most influential musical works of the twentieth century – a bold experiment in new realms of metre, rhythm, tonality and dissonance.

The edginess of the tritone, or the flat fifth as it became known in popular music, made it a favourite interval in both blues and jazz. The blues is based mainly on notes from pentatonic (five-note) scales, but these are supplemented by so-called blue notes, most commonly the flat fifth, which are used in passing. In jazz, the flat fifth is ubiquitous, although again it's used subtly to give a constant urge to resolve tension that's at the heart of all jazz playing.

When rock came along, especially in its more muscular forms, the tritone and other dissonances found a natural new home. You hear the tritone at the start of Jimi Hendrix's 'Purple Haze' and virtually all the way through Black Sabbath's eponymous song off their 1970 *Paranoid* album. The band's guitarist Tony Iommi didn't know much musical theory or anything of the history of the tritone when he started out. He hit on the interval simply by experimenting with sounds that evoked a sense of foreboding and doom. Iommi has the bass G jumping to the same note an octave higher before settling back on the C sharp midway between the two and just hanging there unresolved. And as a nice extra touch he does a trill or vibrato on the C sharp, bending the string back and forth, and pulling the note away from a pure tone to give it added, menacing dissonance.

String bending on guitars, and pitch bending on electronic instruments, is a method of sliding smoothly from one note to another. These techniques take us beyond the musical landscape of the twelve fixed points of the equal temperament scale, which we know so well, into the less familiar realm of microtones.

CHAPTER 12

Microtonal Magic

DOLORES CATHERINO IS an Alaska-based composer and multi-instrumentalist with a highly unconventional approach to both her compositions and the instruments on which she plays them. Catherino is an explorer in the realm of microtonality. Her avant-garde musical creations use not the familiar twelve notes upon which our conventional scales are based but instead an octave that's divided into 106 equal divisions, which she calls 'pitch palettes'. Some of the keyboards on which she brings her sonic endeavours to life are custom-designed arrays of buttons, lit up in a rainbow of colours, which would look more at home on a fictional starship than in a music studio. Catherino is one of a growing number of composer-musicians for whom the available selection of tones normally on offer is simply too limiting.[1]

In typical Western music, the smallest interval that we deal with is a semitone, a half step. But there are numerous pitches between any two notes separated by a semitone. For example, exactly midway between middle C, with a frequency of 261.6 Hz, and C sharp, at 277.2 Hz, is a note with a frequency of 269.4 Hz. This is called C half sharp. It's half a semitone, or one quarter tone, higher than middle C and is a perfectly valid note. So why don't we ever find it used? The fact is, we can see it used, along with many other notes that lie between the familiar ones of the 12-TET scale, if we venture into the strange and alluring world of microtonal music.

The word 'microtone' seems to have first emerged early in the twentieth century. English musicologist and critic Arthur Strangways used it in a written piece in 1914, although Irish musician Maud MacCarthy claims to have coined the term before 1912 to avoid the mistake of referring to the *shrutis* of Indian music as quarter tones. MacCarthy, also known as Omananda Puri, was a talented violinist who'd toured with the Boston Symphony Orchestra before neuritis put an early end to her playing career. During her time on the subcontinent she became steeped in Indian music and mysticism. Later, she campaigned for the harmonium to be abolished in Indian music, arguing that, being capable of playing only certain fixed pitches, it was incompatible with the melismatic or fluid singing style characteristic of the East. In her view, keyboard and other fixed-pitch instruments, with their rigid notes, had led, in the West, to a decline in vocal skills.

We've come to think of microtonality as being somehow outlandish when, in fact, it's common in non-Western cultures. In traditional Indian music the term *shruti* or *śruti* describes the smallest interval of pitch that the human ear can detect. An octave, in Indian theory, is divided into twenty-two *shrutis*. The divisions aren't precisely equal as they are in our current Western system of 12-TET, but the nearest comparison would be to chop each semitone in half so that there are twenty-four such 'quarter tones' to the octave. But though this is tempting, as Maud MacCarthy pointed out, it's also misleading. In practice, the microtonal intervals in classical Indian music vary depending on the style and tradition being performed, and the octave isn't divided evenly as it is in equal temperament.

The concept of *shruti* is deeply rooted in Indian music theory, with early references found in the ancient Vedic texts. The *Natya Shastra*, a foundational text on music, dance and drama, also discusses the importance of *shrutis* in creating the aesthetic experience of *rasa* (emotion). Anyone listening to a singer of classical Indian music will be aware of how different the sound is from anything with which we're familiar. Unlike the Western system, based on equal temperament, *shrutis* enable a more fluid and organic melodic progression. Even without a technical understanding of what we're hearing, there's the sense of something more flowing and subtle, capable of finer forms of melody and emotional expression.

The *Natya Shastra*, a Sanskrit treatise on the performing arts, dates back two thousand years or more. Its six thousand poetic verses cover everything from the structure of a play – even the construction of a stage – to body movements, make-up and costumes, musical scales and instruments, and how to integrate music with a dramatic performance. At the heart of all these teachings is the *shruti*, the atomic building block of the Indian musical world and a concept that's referred to in the even earlier Vedic scripts. Entertainment, asserts the *Natya Shastra*, while a desired effect of performance arts, isn't the main aim. The ultimate goal is nothing less than to transport each member of the audience to a higher state of reality in which they experience the essence of their own consciousness and reflect on deep spiritual and moral questions. Central to this quest and necessary to create the aesthetic experience of *rasa* – the 'taste' or 'essence' of the performance – is the *shruti*.

Shrutis, from the Sanskrit meaning 'to hear', are the subtle pitch elements that exist between the primary notes or swaras. In the Western system, we talk about semitones and scale notes. In Indian classical music, while there are seven primary swaras, there are twenty-two *shrutis* within an octave, allowing for a highly nuanced melodic expression. The intervals between swaras aren't fixed but can depend on the style and tradition of whatever's being performed. The melodic framework of a piece is known as the *raga* (literally 'colouring' or 'dyeing'). The closest comparison is with the different Western musical modes. But in reality, *raga* is unique to the Indian tradition: an array of melodic structures and motifs with the goal of engaging the emotions of the audience in a specific way. A *raga* is notated like the score of a Western work but provides a musical framework within which to improvise.

Although the microtonal intervals can vary depending on the *raga*, the number of *shrutis* separating the principal notes, or swaras, is typically Sa (4), Re (3), Ga (2), Ma (4), Pa (4), Dha (3) and Ni (2). The abbreviations (similar to our 'do, re, mi') are for Shadja, Rishabh, Gandhar and so on.

Western singers find some intervals harder to pitch than others. The tritone – an ascending augmented fourth or descending diminished fifth – can be tricky, as can a descending major seventh. But these are nothing compared with the difficulty of mastering the *shrutis*. And master them you must in order to capture the essence of a *raga* and evoke the intended emotion or mood. While the concept of *shruti* is most prominent in the vocal tradition, it's also applied to various stringed instruments, such as the sitar and sarangi, which have movable frets or continuous fingerboards to enable microtonal adjustments.

So deeply embedded are *raga* in the Indian psyche that it's no surprise to find that numerous tales have sprung up in connection with them. 'Raga Kedar', for instance, is notoriously difficult to perform and, by some accounts, even dangerous because of the amount of thermal energy it's said to generate. Legend has it that the great sixteenth-century Hindustani musician Tansen almost died through heat exhaustion while performing it.

We naturally think of India in connection with microtones and assume that microtonal music always involves very small intervals. But in a broader sense, it can refer to any tuning that differs from Western 12-TET. Under this umbrella comes Indonesian gamelan, traditional Thai, Burmese and African music, and any tuning system that involves just intonation, meantone temperament or other alternative tuning. The Russian composer Ivan Wyschnegradsky introduced the term 'ultrachromatic' for intervals smaller than a semitone and 'infrachromatic' for those larger than a semitone.

In the West, we might be inclined to think that microtones in our musical tradition are a bit radical and experimental – a new toy of avant-garde composers and edgy rock musicians. But independently of what was happening in India, the ancient Greeks were quite familiar with intervals that were microtonal in nature. The Greeks constructed their scales by dividing and combining tetrachords – four notes separated by three intervals – and in the process created three distinct classes: the diatonic, the chromatic and the enharmonic. The last of these featured intervals that sometimes spanned less than half a semitone and which Aristoxenus called 'diesis'.

As we saw earlier, although the enharmonic scales never caught on in the West, they found a natural new home in the Middle East where mellifluous melodies were preferred over the evolution of harmony and multiple vocal and instrumental ensembles. But that's not to say that microtonality didn't have its moment, even during the Renaissance. Guillaume Costeley, court organist to Charles IX of France, was already fond of using unusual melodic intervals, such as the diminished third and augmented fourth, but in his chanson *Seigneur Dieu ta pitié* ('Lord God Your Mercy') of 1558 he did something even more radical: he specifies that to play the piece calls for an instrument to be tuned in 'thirds of a tone', amounting to nineteen keys per octave. It isn't clear how many times *Seigneur Dieu ta pitié* was actually performed in its day, although with a modern synthesiser keyboard, easily adjustable to nineteen equal temperament tuning, the task would be a lot easier.

Around the same time, Italian Renaissance composer Nicola Vicentino was also experimenting with microtonal intervals. But in his case, he actually built an instrument, which he called the archicembalo, to bring his esoteric creations to life. The archicembalo had two manuals like a harpsichord but differed in that the two sets of keys were there to provide not differences in tone but extra pitches. Both manuals contained all of the usual white and black keys, but in addition each black key was divided into two parts so that a distinction could be made between a sharp or flat note. The lower manual also included black keys between B and C, and between E and F. In total, thirty-six keys were available in any octave, each of which was tuned to a different pitch.[2]

But these examples of 500-year-old European microtonality were outliers. Not until the turn of the twentieth century did composers begin to take seriously the prospect of exploring beyond the standard division of the octave into twelve notes. Ivan Wyschnegradsky and Charles Ives were among the earliest in the field, dividing the octave more finely – typically slicing it into twenty-four divisions, each a quarter tone wide. But this presents a problem when it comes to playing. It's easy enough to specify notes that lie at the midpoint of a semitone interval, but how on Earth can you play quarter tones on an instrument, like the piano, which offers a choice of only twelve equally spaced semitones? The answer, it turns out, is simple (at least in principle): you use two pianos – one tuned a quarter tone lower than the other so that, in combination, they can cover all twenty-four quarter tones per octave. All that's needed then are two accomplished pianists who can work side by side as if they were one – reading from the same score but each playing only those notes that can be accessed on their instrument.

In his Prelude no. 1, Wyschnegradsky repeatedly runs up and down the closely spaced notes of his 24-TET composition, producing the slightly unsettling sounds of an 'ultrachromatic' scale. In 'Three Pieces in Quartertones', by Charles Ives, the focus of the music shifts rapidly back and forth between the pair of pianos, moving like a pendulum between tuning in 12-TET and 24-TET. In the brief moments in which only one of the pianos is playing, we hear what appears to be a sequence of notes in standard 12-TET tuning. But when the other piano re-enters and the two overlap, we're immersed once more in the depths of an alien microtonal sea. Looking at a score for this piece you won't see quarter-tone accidentals – semisharps or semiflats – because each piano by itself doesn't play any microtonal intervals. Rather, it's the interaction between the two instruments, tuned a quarter tone apart, which does the microtonal trick.

Much more familiar to most people than these obscure forays into microtonality are the off-piste tones that are integral to rock music and its most influential predecessor, the blues. A music form that took shape among African Americans in the Deep South of the United States around the 1860s, it incorporated spirituals, work songs and field hollers. Distilled into its sound is the pain and long suffering of slavery, often infused with some dry humour, which finds expression in the characteristic 'blue' notes – chromatic or microtonal flattenings of the third, fifth and seventh notes of a diatonic major scale.

There's no more versatile instrument than the human voice for pitching any note and then smoothly sliding it up or down into another. People sang the blues as they worked, or remembered the times when their ancestors worked, on the plantation. Then some of the blues singers added the guitar because a guitar string, like the human voice, can sound a note and then be pushed sideways to raise that note by a semitone or by any fraction of a semitone. Other instruments can execute similar glissandos. At the beginning of George Gershwin's 'Rhapsody in Blue', the clarinet rises seamlessly from a low F to a high B flat. If the clarinet were restricted to the 12-TET system, as the piano is, the beginning of this piece would sound very different. But since it can access notes between the standard dozen pitches, it can achieve a gloriously smooth, soaring effect.

Blue notes add a sense of tension and emotional edginess to a song. It's not surprising they found their way into the seminal rock music of the 1960s. Although most pop and rock sticks to the standard 12-TET system and notes that you can find on a piano, vocalists and guitarists of these genres aren't shy to bend their tones for extra effect and drama.

In electronic pop, a smooth upward shift in tone, passing through a continuous sequence of pitches, is called a riser. There's one immediately before the chorus of 'Uptown Funk', a song performed by Bruno Mars on record producer Mark Ronson's fourth album, and it's a common ploy used in electronic dance music to create a sense of anticipation.

Incidentally, a fascinating effect can be achieved by combining continuously rising tones with some well-timed volume fades. One way to demonstrate it is to start out with four notes an octave apart – say, A4 (concert pitch), A3, A2 and A1 – each consisting of a pure sine tone. Now, keep the middle two constant but make the top and bottom pitches both rise by an octave over the same interval of time. As the high note rises, steadily decrease its volume, while as the low note rises, steadily increase its volume. Loop around to the start if you want it to continue indefinitely. The result is a sound that appears to be getting higher and higher all the time but never actually goes anywhere. It's called a Shepard tone after the American cognitive scientist Roger Shepard.

Some bands have gone out of their way to incorporate microtonality into their music. In 2017, the Australian band King Gizzard and the Lizard Wizard released their album *Flying Microtonal Banana*, written entirely in 24-TET. King Gizzard's guitarist, Stu Mackenzie, wrote the songs on the album using his baglama, a Turkish instrument with frets at intervals that resemble Western quarter tones. To play their music live, the band use guitars with extra frets on the neck to access the required microtones. The resulting tonality has the feel of old and new, East and West – rock with an exotic twist. King Gizzard's sound focuses, too, more on melodic development than harmonic development. Good-sounding chords are hard to construct using microtonal intervals – far harder than simply stacking thirds, fourths, fifths and the like. Too much in the way of harmonised microtones might also threaten to overwhelm the listener.[3]

Refretted microtonal guitars of the American classical guitarist John Schneider.

Radiohead – no strangers to sonic experimentation – have added microtonality into their mix. Guitarist Jonny Greenwood orchestrated a microtonal string arrangement for the song 'How To Disappear Completely' on the band's fourth album,

Kid A, released in 2000. The strings move randomly between pitches, creating a chaotic and unnerving texture. It's difficult to identify clearly the pitches played by the strings – and that's kind of the point. They spend as much time in the spaces between the standard tones as they spend in them, resulting in an ultrachromatic and tonally disorienting sound.

Our conventional system of 12-TET is based on the idea that the octave is the only interval that's strictly in tune – perfectly consonant. The rest of the equally spaced intervals are all a bit dissonant. Generally, the minor intervals are either flat or sharp – even the 'perfect' fourth and fifth intervals are slightly flat and slightly sharp, respectively. But as mentioned earlier, the octave interval itself can seem not entirely pure – a note high on a piano keyboard, for example, sounding a tad flat compared with exactly the same note several octaves lower. To put it mathematically, whereas an octave is defined by a linear relationship – jump an octave and you double the frequency – the perception of pitch appears to be non-linear. Our entire conventional system of tuning in the West is based on the assumption that because an octave interval is consonant in physics it will also sound consonant when perceived by us. But the assumption is flawed.

Musicians and composers like Dolores Catherino, mentioned earlier, are using electronic technology to pioneer polychromatic musical scales with high-pitch resolution so that notes, intervals and complex harmonies can be tailored to how our hearing actually works. In this way, pitch becomes less of a fixed quantity and more of a dynamic process that puts the listener back at the focus of the music.

In time, we may come to look back on 12-TET as an important – but somewhat coarse and primitive – compromise. It enabled easy modulation between chromatic keys but, like fast food over fine dining, it starved us of the subtleties that our senses are capable of discerning. Equal temperament, in which all the intervals between adjacent notes are the same, creates a mismatch between the tempered pitches and the embedded harmonics. We've become so accustomed to a relentless diet of 12-TET that we no longer notice that almost every interval played is slightly dissonant and that the harmonics that inevitably arise from tempering are out of balance.

Historically, there was no way around this except by using *unequal* temperaments. These had the advantage of retaining distinct moods and 'colours' for each key and purer harmonic intonation, but they became increasingly impractical given the way music in the West developed.

Now, thanks to electronic instruments and the introduction of MIDI, we have the opportunity to expand our musical experience by embracing microtonality – not as some occasional, exotic spice but as the new norm. Programmable microtonal instruments make it possible to go beyond simple, linear tuning to a much higher-resolution, non-linear, 'polychromatic' (to use Catherino's term) paradigm. The traditional closed circle of fifths, and rigid demands of the octave, might become a thing of the past. In its place would be open, continuous spirals of pitches and multidimensional harmonic combinations, with interval spacings finely tuned not according to the dictates of musical convenience but the inner workings of the human mind.

CHAPTER 13

Songs of the Cosmos

TWO GOLD-PLATED COPPER phonograph records, each containing ninety minutes of music from different ages and regions of the world, are now in interstellar space bound on endless journeys to the stars. They're fixed to the sides of the *Voyager 1* and 2 spacecraft, which were launched from Earth in 1977. Each record comes complete with a stylus and pictorial instructions so that, if they're ever found by inquisitive aliens, it will be possible to play them and learn about their creators.

The *Voyager* probes were designed and built to take advantage of an alignment of the outer planets that happens only once every 175 years. At these special times, all four big planets in the solar system – Jupiter, Saturn, Uranus and Neptune – line up in such a way that a single spacecraft can visit each of them in turn and, during the fly-by, get a 'gravity assist' to put it on course for the next encounter. The beauty of this kind of manoeuvre is that if a spacecraft approaches on just the right trajectory, its path and speed can be adjusted perfectly to make the next planetary rendezvous for free – without the need to burn precious fuel.

Both *Voyagers*, mission planners knew, would pick up so much speed from skimming past the giant planets that they'd be catapulted beyond the Sun's gravitational grasp and out of the solar system forever. It was decided, therefore, to use them to send a time capsule to the stars – a memento of their place of origin – in the event that some smart ET would come across them in the distant future. That must have seemed an even more remote possibility back in the mid-1970s than it does now, given that not a single exoplanet had yet been discovered.

The *Voyager* Golden Record, its cover and the *Voyager* spacecraft to which it is attached.

It wasn't the first time a message had been placed aboard star-bound craft. The twin *Pioneer* probes, 10 and 11, launched several years earlier, each bore a gold-anodised aluminium plaque etched with a male and a female figure, along with symbols to show Earth's location in space. But the *Voyager* Golden Record contains far more information – both images and sounds – intended to convey the range of life, environments and cultures on Earth. In addition to the ninety minutes of music, the Golden Record features 116 pictures, a variety of natural sounds and spoken greetings in fifty-five languages.

It fell to a committee chaired by planetary astronomer and science communicator Carl Sagan to determine the contents of *Voyager*'s cosmic LP. To help with the decision, the committee included two American ethnomusicologists, Alan Lomax and Robert E. Brown. In the end, extracts from twenty-seven pieces of music were selected to represent the totality of human musical output over the centuries. Among the eclectic mix were a men's house song from New Guinea, an Indian *raga*, two Aborigine songs, 'Morning Star' and 'Devil Bird', a tune played on Peruvian panpipes and drum, Stravinsky's *The Rite of Spring*, pieces by Bach, Beethoven and Mozart, and Chuck Berry's 'Johnny B. Goode'.[1]

As varied as these selections are, we've no trouble in recognising that they *are* all musical. But if clever extraterrestrials ever stumbled across a Golden Record and managed to play it as intended, what would they make of its contents? And, by the same token, if alien music somehow reached our ears, would we appreciate it for what it was?

As I write, *Voyager* 1 is 25 billion kilometres from Earth, making it the most distant human-made object. Radio waves, travelling at the speed of light, take almost a day to cross that vast expanse. The spacecraft is moving seventeen kilometres further from home every second, but even that seemingly breakneck rate of travel is a snail's crawl when faced with the immense gulfs between stars. If *Voyager* 1 were heading directly towards the nearest star to the Sun, Proxima Centauri, which it isn't, it would take about 73,000 years to arrive. In fact, it's travelling in a completely different direction, which won't bring it within a light-year of another star for the next quarter of a million years.

There's no way that we'd be able to detect an inert alien spacecraft, the size of a compact car, if it passed by the Sun a light-year away. But let's assume that our extraterrestrial neighbours, if they exist, are so technologically savvy that they can spot a bit of artificial debris from trillions of kilometres away. They haul in *Voyager 1* for close inspection and manage to extract the sights and sounds from the strange circular disk on its side. What will they make of our musical offerings? Will they appreciate what they are? Is music, in fact, universal?

Mathematics certainly is. Pi – the ratio of the circumference of a circle to its diameter – is the same whether you're from Earth or the fourth planet of a star on the other side of the galaxy. Music, as we've seen, is highly mathematical in its foundations, so wherever music springs up in space or time we'd expect it to have some common features. The perfect fifth, for instance, is a consonant interval wherever and whenever you live. That's simply a fact of maths and physics. And as soon as some basic properties and relationships like this form a common ground, much of the rest of music may start to show similarities, no matter the place and time at which it arises.

From an early age, our brains become attuned to the music that's pervasive around us, just as they become accustomed to the local language, the tastes of our home cuisine and the ways of the people with whom we grow up. Music from other cultures may sound exotic and surprising when we first hear it, and yet, for the most part, it's still recognisable as music. The different scales, intervals, rhythms and structures of musical pieces from other parts of the world may take some getting used to, but we can instantly grasp that they *are* musical. This is because they're all based on acoustic patterns that can be reduced to relatively simple mathematical relationships that govern the basic elements of melody, harmony and tempo.

From the Pythagoreans stemmed the notion that nature itself can produce music – that the planets hum harmoniously as they travel around their orbits. This idea of the harmony of the spheres, or *musica universalis*, may seem quaint nowadays, but ancient theories have a habit of popping up in new guises. Atomism was reborn as modern atomic theory; quintessence is one of the possibilities for explaining dark energy. And astronomers have recently made a remarkable discovery about the planetary system of a star called HD 110067, which lies 105 light-years away in the constellation Coma Berenices. The six planets of this system are all similar in size, ranging from two to three times the diameter of Earth. They're of a type known as sub-Neptunes. But the most fascinating thing about them is not the similarity of their appearance but the finely tuned choreography of their orbital dance. In the time it takes for the innermost planet, *a*, to complete three orbits, the next planet out, *b*, makes two revolutions. For every three revolutions of *b*, the third planet, *c*, makes two. This pattern repeats for planets *c* and *d*. Then there's a change so that the fifth planet, *e*, goes around its orbit three times for every four of *d*, and similarly the 4:3 pattern is repeated for the outer two worlds, *e* and *f*. Combined, this means that the innermost planet of HD 110067 completes exactly six orbits for every one of planet *f*.

Frequency ratios of 3:2 and 4:3, as we've seen, are familiar and important in music, corresponding to the intervals of a perfect fifth and a perfect fourth. You might think it extraordinary that these same ratios would crop up in the motions of a natural system, but in fact they're a direct consequence of a phenomenon known as orbital resonance. Closer to home, three of the large moons of Jupiter – Ganymede, Europa and Io – dance to the same kind of tune. Io orbits Jupiter in exactly twice the time it takes Europa, which, in turn, takes twice as long as Ganymede. The 4:2:1 ratio comes about because of the mutual gravitational interactions of the three moons which act like the carefully timed pushes given to a child's swing to keep it moving to the same height with every to and fro.

Among the planets of the solar system there's only one notable resonance: between Neptune and Pluto (although the latter has now been demoted to 'dwarf planet' status). Neptune makes three trips around the Sun for every two of Pluto. The mutual gravitational nudges between these worlds – dominated by Neptune, which is eight thousand times more massive than Pluto – are such that they're locked indefinitely in a synchronised pas de deux. One of the reasons for the stability of the resonance is that Neptune and Pluto never approach each other more closely than twenty times the Earth-to-Sun distance, so the interactions are subtle and self-correcting rather than disruptive.

There were too many disturbances, including collisions and close encounters, during the early days of the Solar System for any other planets to have settled into resonant orbits. But the planetary system of HD 110067 is different. Its evenly sized worlds occupy the same resonant orbits now as they did when they were first formed billions of years ago. A rare one percent of planetary systems in the galaxy are thought to show this kind of mutually synchronised rhythm.

If only the worlds of HD 110067 each sounded a note tuned according to their orbital periods. Then there'd be an actual case of harmony of the spheres – an ethereal six-note chord composed of fourths and fifths. In the absence of humming planets and of sounds that can travel through a vacuum, we're left to add our own musical accompaniment, which is exactly what researchers have done. They've created a musical piece with notes that play each time a planet in a sped-up simulation of the resonance chain completes one orbit and rhythms that correspond to their orbital periods.[2]

Astronomers have found several other planetary systems, in our neck of the galactic woods, which display a similar kind of mutual resonance. A mere forty-one light-years away is the dim red dwarf TRAPPIST-1 with its entourage of seven planets, at least two of which are similar in size to Earth. The orbital periods of the TRAPPIST septet range from just one and a half days to nineteen days and have ratios of 8:5, 5:3, 3:2, 3:2, 4:3 and 3:2 between neighbouring pairs, moving from inner to outer. In terms of musical intervals these correspond to a minor sixth, a major sixth, two perfect fifths, a perfect fourth and another perfect fifth.

Any alien astronomers equipped with instruments at least as powerful as those on Earth will have discovered resonant planetary systems like HD 110067 and TRAPPIST-1. They'll know how common simple frequency ratios are in nature, whether they're associated with gravitational interactions or consonant sounds. This connection between maths and physical phenomena has nothing to do with local circumstances or culture. It's woven into the very fabric of the universe. It's no coincidence that the most harmonious intervals of music involve the same simple fractions as those between the orbital periods of neighbouring worlds in resonant systems of planets.

Because music is fundamentally mathematical, and because maths is universal, it seems likely that if other intelligent species have evolved elsewhere in the galaxy and beyond, they too will have come up with music of some form. The variety is likely to be immense, just as it is on Earth. Think of a spectrum that embraces Gregorian chant, flamenco, bluegrass, gamelan, Noh, fusion, psychedelic rock, Romantic classical and all other types of music from around the world and across the ages. Now add in the possibility of new genres of which human minds have never conceived, and the extent of what might comprise music across the cosmos becomes clear.

Anatomy comes into play. Our appreciation of music is limited by the frequency range to which our ears are sensitive – roughly 20 to 20,000 Hz. Other animals can hear well beyond this range: down to 0.05 Hz in the case of pigeons and up to about 200,000 Hz in some species of bat. Wolves have triangular-shaped ears that can rotate independently and pinpoint the direction of sounds from several miles away. Elephants can sense vibrations of the ground through their feet. Possibly the processing power of some extraterrestrials would be vastly greater than that of our own brains, or that of our fastest computers, such that they could appreciate some complex sounds as being musical that would, in a sense, go over our heads. In theory, there's no limit to the types of sounds alien anatomies and intellects might be capable of handling, in terms of frequency, amplitude, ability to discern differences in pitch, tempo and the like, or any other acoustic parameter.

Environment, too, is a factor in what can be heard. In underwater environments, for instance, things sound different. When sound waves travel through air they cause our eardrums to vibrate. The vibrations are transmitted to three tiny bones, called ossicles, in the middle ear and then to sensory cells in the inner ear, which fire off electrical signals to the brain. We've evolved to hear well on land so, not surprisingly, our underwater hearing isn't nearly as good. Much of it takes place through bone conduction when vibrations in parts of the skull are transmitted directly to the inner ear.

Bone conduction happens all the time, but we're not usually aware of it because most sounds in air, especially at higher frequencies, enter via the eardrum and ear canal. Plug your ears and bone conduction is the reason you can still hear your own voice. It also explains why our voice sounds different to us when it's recorded and played back than when we speak directly. Because the skull is better at conducting lower frequencies than higher ones, we tend to perceive our voice to be lower and fuller than others do, and are surprised when recordings of it make it sound higher and lighter than what we're used to. Bone-anchored hearing aids are a treatment option for some types of hearing impairment, and bone-conduction headphones with speakers that strap over the mastoid process – the dome-shaped bone behind the ear – are useful for some activities like scuba diving. Musicians, too, can benefit from this alternative way of perceiving sound.

Beethoven was just twenty-eight when he noticed he was losing his hearing. But it didn't stop him from composing some of the world's greatest music. Between 1803 and 1812, when he was almost completely deaf, he wrote an opera, six symphonies, four solo concerti, five string quartets, six string sonatas, seven piano sonatas, five sets of piano variations, four overtures, four trios, two sextets and seventy-two songs. The cause of his deafness isn't known, though post-mortem analysis of his skull and a lock of his hair suggests one possibility: the specimens contain an abnormally high level of lead, and one of the side effects of lead poisoning is hearing loss. Beethoven went through phases of heavy drinking, and cheap wines of the time were often sweetened with lead acetate to improve their flavour. He also drank from a lead goblet.[3]

Being a musical genius, Beethoven could presumably conjure up the accurate pitch of notes, and the sound of complex combinations of notes, in his head. But it's said he had another tool at his disposal. When composing at the piano, he'd clench a wooden stick between his teeth, which was attached at the other end to the piano. The vibrations from the instrument would thereby be transmitted to his jaw, and then, by bone conduction, to his inner ear, giving him some physical perception of the sounds he was playing.

We don't hear well underwater because it's not our natural environment. But in fact, sound is transmitted much better through water than through air and, depending on its frequency, can propagate over several kilometres. Visibility in the ocean, by contrast, is generally poor – at most thirty metres or so – which explains why many marine mammals have poor eyesight but well-developed hearing and complex means of sound communication. When our air-adapted ears are submerged, we can't tell from which direction sounds are coming and, if those sounds are quiet, struggle to pick them up at all. Aquatic mammals, on the other hand, have auditory anatomies ideally suited to a watery soundscape. In dolphins, porpoises and other toothed whales, the ear canal is completely closed off when underwater: it simply doesn't work. Instead, sounds enter the head through the lower jaw, which is filled with a special type of fat that conducts sound incredibly well. Vibrations travel along this mandibular fat channel directly to the middle ear, which is encased in a bony shell known as the acoustic bulla. The bone of the bulla is among the densest and hardest materials known in the animal kingdom, making it ideally suited for transmitting sound from the fat channel to the ossicles of the middle ear.

Auditory fat also fills the structure known as a melon in the bulbous foreheads of dolphins and their cousins. The melon acts like a lens to concentrate incoming sounds as well as to focus and modulate the animal's vocalisations. It plays an essential role in both communication and echolocation. Dolphins signal to one another using a variety of whistles and ultrasonic clicks. They're also attracted by music played to them underwater through a hydrophone, especially if it's high-pitched as on a flute or piccolo, or sung by a soprano. Whether dolphins and whales make music of their own or not, it certainly seems they have an innate musical sense.

Are there entire underwater civilisations on other worlds playing compositions by aquatic Bachs and Beatles? It seems that subsurface oceans might be common throughout space, following the discovery that they probably exist on several moons in the solar system, including Jupiter's Europa, Ganymede and Callisto, Saturn's Titan and Enceladus, and

Neptune's Triton. But if a marine alien race on some distant waterworld did create its own musical works, they'd have to be played on very different instruments to those with which we're familiar. An inkling of the possibilities comes from the work of a Danish ensemble called Between Music. The five members of the group carried out experiments with the help of divers, instrument makers and scientists to develop effective ways of making music underwater and now give live performances of their compositions from inside large aquarium tanks.[4]

Imagine the challenge of creating subaqueous instruments. Water is a much denser medium than air, so it absorbs and dampens vibrations. And sound travels about four times faster underwater, making the sound waves four times longer. Materials have to be selected with care because some will corrode, get sodden or dissolve when submerged. Ferrous substances and wood, for instance, need to be avoided.

Among the remarkable instruments played by Between Music are a hydraulophone – an underwater organ first developed by Canadian scientist and inventor Steve Mann in 1985. The hydraulophone uses water, rather than air, to produce sound by manipulating water jets through a variable system of tubes and flow vessels. Mann's colleague, Ryan Janzen, was commissioned by Between Music to design a new addition to the hydraulophone family. American software developer and inventor Andy Cavatorta re-engineered a glass armonica, or crystallophone, such that its structure and acoustic properties were tailored to a liquid medium. Cavatorta was also asked to create a new underwater string instrument based on the hurdy gurdy. Christened a rotacorda, it produces sound when a crank-turned wheel rubs against a series of strings. A carbon-fibre violin, a carefully chosen set of Asian singing bowls and various items of tuned percussion complete the instrument collection. The team has also perfected a distinctive, haunting vocal technique for underwater singing. Because the artists have no breathing apparatus, the timing for getting air is often written into the composition. Equally important is the precise placement of instruments and microphones, as well as keeping the water at the right temperature so that the instruments play at the desired pitch.[5]

If humans can adapt to making music underwater, it's reasonable to assume that intelligent aliens to whom this would be a natural environment could do the same, and even more effectively. The same is true of a musically inclined civilisation that evolved in an atmosphere different in make-up and pressure to our own. Music is possible in any medium through which sound can travel, though its character and quality will be influenced by physical properties like density and pressure.

Perhaps music is possible even without sound or hearing. In 'The Secret Sense', a 1941 short story by Isaac Asimov, Lincoln Fields, a rich American living on Mars, learns of a sense, unknown to humans, that Martians possess. He insists on experiencing it but is warned that the effect of the injection, which will activate the sense in him, will last for only a few minutes, after which it can never be repeated. Fields goes ahead with the process and is consumed by the wonders of a Martian musical instrument, the *portwen*, as colours, sounds and odours merge and transform into a new, indescribable mode of sensation. When it ends, he's devastated by the knowledge that he'll never possess the 'secret sense' again.

Music, like life, may take many forms across the immensity of time and space. Which of the pieces included on the *Voyager* Golden Record would most obviously sound musical to alien ears? Some believe it would be the works of Bach, that most mathematical of composers. In fact, of the twenty-seven selections contained on the Record, three are by Bach – extracts from the *Brandenburg Concerto* no. 2 in F, the 'Gavotte en Rondeau' from the Partita no. 3 in E major for Violin, and the Prelude and Fugue no. 1 in C Major from *The Well-Tempered Clavier*, Book 2. The Bach contributions last twelve minutes and twenty-three seconds, or roughly one-seventh of the playing time of the whole record, reflecting the belief of those who assembled the collection that the highly structured nature of Bach's pieces, including his clever and complex use of counterpoint to interweave multiple melodic lines, would appeal to both the intellect and aesthetics of any advanced beings who came across the spacecraft.

It's been argued that music may be the best way to open an extraterrestrial dialogue. In the movie *Close Encounters of the Third Kind* the aliens play a five-note sequence from a major scale as a greeting – re mi do do (down an octave) so – evidently as a non-verbal means of establishing 'first contact'.

This idea of music as a cosmic lingua franca isn't new. In the seventeenth century, English clergyman Francis Godwin, Bishop of Hereford, wrote a short book called *The Man in the Moone* (published posthumously in 1638) in which his intrepid astronaut Domingo Gonsales encounters a race of Lunarians who communicate via a musical language. Godwin's idea built on a description of the spoken Chinese language, with its tonal sounds, by Jesuit missionaries who had recently returned to Europe. In Godwin's tale, the

Lunarians used different notes to represent the letters of their alphabet.

In the 1960s, the German radio astronomer Sebastian von Hoerner, who wrote extensively on the search for extraterrestrial intelligence (SETI), argued in favour of music as a medium of choice for interstellar communication. It might well be, he suggested, that alien music would share some features in common with our own. Wherever polyphonic music evolved there'd be only a limited number of workable ways to produce harmonious sounds. To allow for modulations from one key to another, an octave has to be divided into equal parts and the corresponding tones have to be at frequencies that bear certain mathematical ratios to one another. The compromise that's emerged in Western music is the 12-TET scale. That scale, said von Hoerner, might crop up in the music of other worlds, as might a couple of other equal temperament scales that offer good compromises for polyphony: the five-tone scale and the thirty-one-tone scale. The latter was written about in the seventeenth century by a number of scholars, including astronomer Christiaan Huygens, and might be the scale of choice among beings whose auditory systems are more sensitive than our own. On the other hand, aliens whose biology made them less adept at differentiating between closely spaced pitches would perhaps be more likely to use 5-TET.

It's often assumed that the first message we receive from the stars will be scientific or mathematical in content. But what better way to extend a greeting than by sending a really good piece of music, one that hasn't just a logical basis but is full of the passion of its creators. After all, it's well known that music – perhaps more than any other mode of expression – has a remarkable capacity to evoke emotion and bring people together. It has the power to build bridges between different cultures and languages, forging links between people who might not otherwise meet or agree on other things. Just listen to crowds of strangers from different places and backgrounds singing in unison at concerts and festivals, and music's profound ability to unite becomes clear.

Scientists have been beaming radio messages to the stars since 1974, when a powerful signal was sent from the giant Arecibo telescope, in Puerto Rico, towards the Hercules Globular Cluster, M13. In 2001, the first cosmic transmissions with musical content were directed towards several nearby Sun-like stars from the Yevpatoria radio telescope in Crimea. Astronomer Alexander Zaitsev, who supervised the project, believed that art should be central to any attempt at interstellar communication because of its emotional and intellectual content. Seven songs, by Beethoven, Vivaldi, Gershwin and others, were selected by students and played on the theremin by three performers in Moscow. The renditions were then broadcast as the *First Theremin Concert for Extraterrestrials* from the Yevpatoria dish, which had been specially modified to accept input from the instrument. Seven years later, using one of the dishes of its Deep Space Network (DSN), NASA fired the Beatles' song 'Across the Universe' in the general direction of Polaris, the North Star, to mark several anniversaries – the fortieth of the song's recording, the forty-fifth of the DSN and the fiftieth of NASA itself.

Music is the consummate fusion of passion and logic, the yin and yang that all intelligent beings across the universe may experience. Even if an extraterrestrial lacks the ability to hear sound in the way we do, it would be able to access the central elements of music – pitch, rhythm, harmony and so on – by studying the way these elements are encoded in radio waves, either as an analogue or digital signal. Within the fields of SETI and METI (messaging extraterrestrial intelligence), enthusiasm for sending music to the stars has grown over the years.

Sónar is a three-day festival held every year in Barcelona to celebrate electronic music, art and design. For its twenty-fifth anniversary in 2018, the festival partnered with METI International, a non-profit research organisation founded by American astrobiologist and SETI researcher Doug Vakoch, and the Catalonia Institute for Space Studies, to send out a series of interstellar messages using a radar system in Tromsø, Norway. The target was Luyten's star, a red dwarf about twelve light-years away around which orbits the closest known potentially habitable planet outside of the Solar System. Each transmission began with a kind of mathematical 'hi' – the first thirty-three prime numbers repeated on two alternating radio frequencies to announce that the source isn't natural. As for the music itself, consisting of a series of 'songs', each just a few seconds long, it won't be winning a Grammy anytime soon. One of the tunes is based on the atomic numbers of several common elements, such as oxygen and silicon, converted into pure-tone frequencies. It's not an easy listen but is rich in information – and that's the point. The compositions are meant to convey something about us in a form that's partly artistic. And in this respect, they represent something new: an attempt to communicate with extraterrestrials through music designed specifically for that purpose.

In his book *Underland*, Robert Macfarlane mentions a proposal made in the 1990s for warning far-future generations – effectively aliens compared with our present selves – of the presence of radioactive waste buried deep underground. Why not, it was suggested, build a virtually indestructible aeolian instrument to tune the desert winds of that far-off time to D minor. The chord long held to convey sadness and despair would serve as a warning that danger lay nearby.

Presently a bigger, more sophisticated effort than the one organised for Sónar is underway to create a musical composition intentionally designed for interstellar transmission. Called the Earthling Project, it's the brainchild of the SETI Institute's artist-in-residence program. The idea is to crowdsource original music from around the globe, and then, using this as a basis, compose something specifically for sending to the stars.

The argument goes that no existing pieces of music can truly represent all of humanity, because such pieces will always be culturally biased. To get around this problem, people from all over the world will be able to contribute to a new musical message designed specifically for alien ears. A web-based platform will allow samples of human voices to be recorded anywhere on the planet. Then the voices will be merged with electronic synth, a live ensemble, samples from musicians (both amateur and professional) and the results of sonification of data from space, to create a seven-part symphony called *Earthling*.[6]

Until now, humans have always made music for themselves or for their deities. But 'Earthling' represents something new: an attempt to connect two different forms of intelligence across the vastness of space. Whether the composition is ever broadcast into the cosmos remains to be seen, but the project is still unique and worthwhile: the first attempt to create a lengthy musical offering for our interstellar neighbours – if any exist.

Meanwhile, on our own planet, a new form of intelligence is arising that may soon far surpass our own. 'AI' is the buzzword of the day. There's AI-created text, images, videos and, of course, music. English music producer and rock musician Steve Wilson was surprised to learn that songs composed and performed, in his style and with his (synthesised) voice, had been generated using AI software. 'Even I', he said, 'really struggle to hear that it's not me singing these songs.' And we're just at the dawn of AI music-making. Already it's possible to type a few lines into an AI music generator, describing the genre of song, the subject matter and the famous singer you want it to sound like, and the software will do the rest. 'A song in the style of ABBA from the 1970s' – no problem. 'Freddy Mercury performing a Queen-like song with a strong beat' – done, in less than a minute.

With AI on hand, who needs real musicians anymore? We're not quite there yet, but in five years or so, you'll be able to generate whatever music your heart desires in real time, save it, share it with your friends or erase it and create more. At that point, how could human musicians survive and continue to make a living? The fact is, as Wilson points out, 'We're in the midst of a seismic change in the way music is made and how people engage with it.' Do we care? Is it important to know whether what we listen to comes from actual people or a bank of electronic processors? And what about live performances – concerts and festivals? It's already possible, as in the case of the ABBA Voyage show, to have virtual performers on stage. The technology is progressing with lightning speed: the only issue is how we choose to adapt to it.

CHAPTER 14

Music on the Brain

Music is stored and processed by the brain in ways that other experiences and activities are not. I know this personally because my mother suffered for more than a decade from Alzheimer's. Even after she lost the ability to count or write, or remember the faces of her own family, she'd still react to songs from her youth and could sing along despite being unable to speak in a normal way. The neural regions that held and responded to music evidently survived even as the disease ravaged other parts of her brain.

Glen Campbell, supported by musical members of his family, was able to continue performing for some time even as his Alzheimer's progressed. In between songs he might be lost for words and have no clue where he was, but as soon as the band started up it would be business as usual. Tony Bennett and Lady Gaga staged a pair of critically acclaimed concerts together at New York's Radio City Music Hall in August 2021 after Bennett's Alzheimer's diagnosis had been made public. Dementia choirs have become increasingly popular as their benefits are recognised. There's something about music that differs from other forms of communication in the way we store and access it neurologically and the way it makes us feel.

Music has been an important part of every human culture, past and present. It can play a role in brain development, learning, mood and overall health. There used to be a popular belief that music is processed mainly in the right hemisphere of the brain, along with art and other creative activities. However, recent findings have shown that this isn't the case – that, in fact, music is perceived and processed *throughout* the brain. From studies of patients with brain damage, it's been found that people who've lost the ability to make sense of written text can still read music. Others who lack the motor coordination to button their own clothes have not lost any previously learned ability to play a musical instrument. Today we know that both listening to music and performing it engage nearly every area of the brain that's been investigated.

A common effect of music is to alter our mood and feelings by stimulating the formation of certain brain chemicals. Movie directors use film scores to elevate our emotions at certain dramatic moments. Think of a typical chase or battle sequence in an action movie: it's the music that makes the scene truly epic.

Our brains respond differently to happy and sad melodies. One study showed that after hearing a short piece of melancholy or downbeat music, participants were more likely to interpret a neutral expression as sad. If the melody was joyful, the neutral expression was perceived as being a happy one.[1] When you listen to music that you enjoy, your brain releases a neurotransmitter called dopamine – a chemical that causes a feeling of satisfaction. When listening to a favourite part of a song, you feel the same sort of pleasure, to varying degrees, as you experience when eating tasty food or having sex. So music can give us good vibes, and, however your day is going, having music on in the background can make it seem better. Pleasing music can also reduce blood flow to the amygdala, a major processing centre for emotions, most notably fear, lower the production of the stress hormone cortisol and release oxytocin, sometimes referred to as the 'love hormone' because it enhances bonding and makes us feel connected to others.[2]

Interestingly, if you're sad, then listening to sad music can sometimes improve your mood as well. You might suppose that the last thing that would help, if you're having a bad day, is to play something that reflects that negativity. Why wouldn't a sad person listen to happy music? The reason is that when you're feeling down or depressed, it's often because you feel misunderstood – that the people around you don't appreciate what you're going through. If you were to listen to happy music while in this state, it would only amplify that feeling of detachment. On the other hand, if you put on the right piece of sad music, you can say to yourself, 'That's exactly how I feel. This musician understands me.' So the sad music turns out to be soothing – a companion to your misery, as it were.[3] In the same way, despite a long-standing public perception that heavy music, such as metal, is aggressive, studies have shown that listening to it can lessen negative emotions. Heavy metal, and similar angry-sounding genres, offer catharsis and help to reduce stress-hormone levels. Against a backdrop of rising mental health disorders, especially among young people, music offers a means of escape, solace, expression and connection.

Tuning in to music can impact an exercise regime. As our body senses that it's tired, it sends signals to the brain to stop for a break. Listening to upbeat music, however, competes for the brain's attention and can help override those signals of fatigue. According to a study published in 2012, cyclists who listen to music require seven percent less oxygen to do the same amount of work as those who pedal in silence. It seems that music can help us not only push through the pain barrier to exercise longer and harder but also use our energy more efficiently.[4]

In the past few decades, neuroscientists have made enormous strides in understanding how our brains work by monitoring them through techniques such as functional magnetic resonance imaging (fMRI), which measures brain activity by detecting changes associated with blood flow. It relies on the fact that cerebral blood flow and neuronal activation are coupled: if an area of the brain is in use, blood flow to that region also increases. When researchers asked participants to listen to music while being monitored, they saw multiple regions light up. When they observed the brains of musicians while playing an instrument, the results were even more spectacular – the whole of the brain burst into action like a firework display.

It seems that while simply listening to music engages a surprising number of different parts of the brain, actually playing an instrument gives the cerebral cortex the equivalent of a full-body workout. Not surprisingly then, just as athletes have greater stamina and more toned muscles than the rest of us, musicians have brains that are better connected and, in some areas, more highly developed than average. This finding has important repercussions for education. A study from 2008 showed that children who had at least three years of instrumental music training performed better than their non-musical counterparts on a variety of tasks.[5] The difference was especially marked in auditory discrimination abilities and fine motor skills. But musically trained youngsters also fared better when it came to vocabulary and non-verbal reasoning. The latter involved understanding and analysing visual information, such as identifying relationships, similarities and differences between shapes, and patterns. Abilities like these seem far removed from musical training, so it's important for educationalists and those with influence over school funding and curricula to recognise how learning to play an instrument, or actively taking part in music in other ways, can help children develop a wide variety of important skills.

Plato said, 'Music is a more potent instrument than any other for education.' Today, music in schools is generally underfunded and neglected. Many educators, however, believe it should be restored to a central place in the curriculum as it was in ancient times and in the quadrivium. Recent research suggests it has multiple benefits, from improving motor skills and creative thinking to boosting performance in other subjects such as maths and language. The neurologist Oliver Sacks wrote, 'In terms of brain development, musical performance is every bit as important educationally as reading or writing.'

In recent decades the role of music in cognitive development has been well established. Studying music and, especially, playing an instrument helps with language processing, memory retention, maths, social-skills development and long-term success in life. A study published in 2018 found that 'structured music lessons significantly enhance children's cognitive abilities – including language-based reasoning, short-term memory, planning and inhibition – leading to improved academic performance'.[6]

It's also reassuring to find, as other research has shown, that it's never too late to reap the benefits from learning an instrument or starting to sing. Seniors who take up the piano or another instrument, or simply sing along with others in a choir, appear to be more resistant to age-related cognitive and memory problems. One reason for this might be the creation of alternative connections in the brain that can compensate for cognitive decline as we get older. Regular choir members report that learning new songs is cognitively stimulating and helps their ability to recall things in general.

Singing in a choir brings other advantages – as I can testify, having been a member of a local community choir, the Noteables in Dundee, for the past decade. It helps forge social bonds, but more than that, according to research conducted at the University of Oxford, it's one of the most effective ways of quickly forging close connections between large groups of people. In a world where an increasing amount of social interaction takes place online through tapping keys and watching screens, community singing, it turns out, is remarkably effective at establishing broad, face-to-face social networks. Increasing evidence suggests that in-person social connections can play a vital role in maintaining health. And music, especially communal singing, is among the most potent ways of building such connections and of making people feel part of a close-knit community.[7]

Being part of a cohesive group has been essential for survival throughout our evolutionary history, but members of a group also face challenges such as conflict over resources and mates. In order to prosper, our ancestors needed ways to keep the group together in spite of these conflicts. The fact that music often occurs in social settings, from religious rituals to football games, suggests that it might be an evolved behaviour for creating community cohesion. In Western societies, music-making is often assumed to be the domain of a talented few, but in reality it's extremely rare for a person to have no latent musical ability. The growth of community choirs, open to anyone, demonstrates these inherent skills and suggests that we may be returning to the roots of communal musical behaviour. In light of mounting concerns about loneliness and isolation, and the increasingly urgent search for solutions, it's interesting and encouraging that people seem to be rediscovering a deep-seated potential to connect with one another through song.

The satisfaction of performing together, with or without an audience, is likely to be associated with activation of the brain's reward system. As mentioned earlier, this includes the dopamine pathway, which effectively keeps people coming back for more. Research has found that people feel more positive after actively singing than they do after passively listening to music or after chatting about positive life events. Improved mood probably comes, in part, directly from the release of positive neurochemicals such as beta-endorphin, dopamine and serotonin. It's also likely to be influenced by changes in our sense of social closeness with others.

The wider, physiological benefits of singing and music in general have long been explored. Music-making exercises the brain as well as the body, but singing is particularly beneficial for improving breathing, posture and muscle tension. Listening to and participating in music has been shown to be effective in pain relief, too, again due to the release of neurochemicals such as beta-endorphin – a natural painkiller responsible for the 'high' experienced after intense exercise. There's also some evidence to suggest that music can play a role in sustaining a healthy immune system by reducing the stress hormone cortisol and boosting the immunoglobin A antibody. It's not surprising to find that music has been used in different cultures throughout history in many healing rituals. Today, music therapy is being applied increasingly in our own culture to treat a range of conditions. It can help reduce the frequency of epileptic seizures and alleviate a number of other neuropsychiatric disorders. The well-structured music of Mozart and Bach has been especially favoured for interventions, though other musical styles have also shown potential.

Music therapists tailor their approach depending on a person's needs. Active interventions involve actually making music of some kind – singing or playing an instrument – or physically engaging in some other way, say, by tapping out a rhythm or dancing. Receptive interventions, on the other hand, have more to do with listening to music, played either by the therapist or from a recording, and then exploring the thoughts and feelings that the sounds evoke. A panoply of illnesses and impairments have been shown to benefit from this kind of treatment, including stroke, Parkinson's disease, traumatic brain injury, autism spectrum disorder, mood disorders, pain (acute and chronic) and problems related to breathing.

Both listening to and creating music are beneficial, but because music affects each of us differently, our music *choices* are important. In other words, one person's idea of a pleasurable musical experience might be another person's noise. This has a direct neurological consequence: research has shown that listening to music you like increases blood flow to the brain more than listening to music you don't. So, not surprisingly, the greatest benefits come from engaging with music you actually enjoy.

Occasionally, instead of being beneficial, music can arise as part of a medical condition. Musical hallucination, or musical tinnitus as it's also known, is when a person hears music in his or her head when none is actually being played. When someone experiences this for the first time, they may be absolutely convinced that the sound is from some external source. Typically, a musical hallucination will consist of a short fragment of a simple melody – often from a tune familiar from childhood – but in other cases it may go on for hours. People with hearing loss sometimes notice that the music in a hallucination is as they remembered it, before their hearing loss made sounds more muffled, quieter or with higher frequencies missing.

Although anyone can experience musical hallucinations, they're more common in women than men and in people who live alone, have some hearing loss and are over the age of sixty. Often there seems to be no underlying medical condition, but hallucinations *can* accompany epilepsy and Alzheimer's disease. They're particularly common in people who have obsessive–compulsive disorder (OCD), in which they experience repetitive, intrusive and distressing thoughts, and feel strong urges to repeatedly perform the same actions. As many as four in ten individuals with OCD will experience musical hallucinations at some time in their life.

You'll often hear people say, 'I can't sing, I'm tone deaf.' Almost always, this isn't true. They may not be able to sing *well*, or in tune, but then again most of us aren't very good at things unless we practise and put some effort in. Speaking personally, I started taking singing lessons about seventeen years ago and it did wonders for my ability to breathe properly and pitch reasonably accurately. More or less anyone can hold a tune once they develop 'muscle memory'. Genuine tone deafness, or amusia – when a person has trouble processing pitch or remembering musical information – is rare and usually either a congenital condition or the result of a brain injury.

At the opposite extreme, absolute, or perfect, pitch is the ability to identify by ear any note at some standard pitch or sing a specified note at will. Fully developed absolute pitch is rare. It appears early in childhood and in many cases is an acute form of memory of sounds of a particular instrument, such as the home piano. Some musicians slowly acquire a degree of absolute pitch, if only for the familiar A4 = 440 Hz or, for example, the open strings of a guitar. The musician-producer, multi-instrumentalist and popular YouTuber Rick Beato, encouraged his young son Dylan, through regular practice, to identify combinations of specific notes. Now, if a complex ten-finger chord is played on the piano – even a chord that's extremely dissonant – Dylan can accurately name every note. Beato believes that such comprehensive ability is possible to acquire only in early childhood.[8]

Most people differentiate pitches by comparing one tone to another. For example, if I give you a certain root note, or 'do', as a starting point, you could probably find the note 'so' that's a fifth above it (as in the first two notes of 'Somewhere Over the Rainbow' from *The Wizard of Oz*). But only about one in ten thousand people in the West can identify notes based purely on the sonic information provided by a single tone.

Chinese speakers, on the other hand, are about nine times more likely to have absolute pitch. Psychologist Diana Deutsch of the University of California, San Diego, hypothesised that this is because those whose native language is 'tonal' – it uses pitch to distinguish the meaning of words – are more likely to have a keener ear for pitch in music as well.

Deutsch tested the pitch-identification skills of eighty-eight students at the Central Conservatory of Music in Beijing and did the same with a cohort of English-speaking students enrolled at Eastman School of Music in Rochester, New York. Her study used thirty-six piano notes spanning a three-octave range, generated accurately to concert pitch by a synthesiser. To minimise the use of relative pitch, all intervals between successive tones were larger than an octave. Deutsch found that the Central Conservatory musicians routinely outperformed their Eastman counterparts. For students who'd begun musical training between the ages of four and five, about sixty percent of the Chinese speakers tested as having absolute pitch, whereas only about fourteen percent of the non-tonal-language (English) speakers did. For those beginning between eight and nine, the figures were forty-two percent of the Chinese group and zero of the English group. This has nothing to do with ethnicity or place of origin. What this research shows is that people can routinely acquire absolute pitch providing they're exposed to fine tonal nuances from an early age.[9]

Our brains are remarkably adept when it comes to processing and storing music. Normally we think of music as originating outside ourselves and then being fed to the brain via the ear and auditory nerve. But researchers have found, astonishingly, that it's possible for music stored in the brain to be accessed and played externally. Scientists at the University of California, Berkeley, recreated a famous Pink Floyd track by eavesdropping on people's brainwaves – the first time a recognisable song was reconstructed from recordings of electrical brain activity.

Robert Knight, a neurologist at the University of California, Berkeley, and fellow researchers analysed brain recordings from twenty-nine patients as they were played an approximately three-minute segment of the song 'Another Brick in the Wall, Part 1', from Floyd's 1979 album *The Wall*. The volunteers' brain activity was detected by placing electrodes directly on the surface of their brains as they underwent surgery for epilepsy. Previous work had focused on decoding activity from the brain's speech motor cortex – an area that controls the fine muscle movements of the lips, jaw, tongue and larynx that form words. This new study took recordings from the brain's auditory regions, where all aspects of sound are processed.[10]

Knight and his team hope that their research could eventually help restore the musicality of natural speech in patients who struggle to communicate because of disabling neurological conditions such as stroke or amyotrophic lateral sclerosis. At the moment, speech-assisting devices can be slow and robotic, like the synthetic voice with which physicist Stephen Hawking spoke in his later years. The hope is that the computer algorithms used to recreate music from brainwaves can be extended to add emotional, rhythmic and melodic aspects to assisted speech so that it can have a natural and nuanced sound.

Strangely enough, the way our brains work when it comes to sound has influenced the seating arrangement of orchestras.

The next time you go to a classical concert, notice that the higher-pitched instruments – violins, flutes, clarinets, trumpets and harps – are on the left from the audience's point of view. On the right are the cellos, double basses, bassoons, timpani and so forth, which have a lower range. That's not arbitrary. The same tends to be the case with choirs: sopranos and tenors to the left, altos and basses to the right.

The reason for this bias has to do with our biology. The left cerebral hemisphere tends to process higher pitches better, whereas the right side of the brain has an advantage when it comes to lower pitches. As is the case with all vertebrate species, there's a left–right crossover: the left side of the brain is wired to the right side of the body, and vice versa. So, the left hemisphere, which is more sensitive to higher pitches, gets its auditory information from the right ear.

You may be thinking, given these facts, that from the audience's perspective an orchestra is set up wrong. Shouldn't the violins and higher-pitched woodwind be to the right of the listeners facing them? But the seating arrangement is the way it is not to benefit the audience but to enable the players on stage to hear more clearly the overall sound of the ensemble in which they're immersed. The left ears of the violinists, for example, are facing the lower-pitched section of the orchestra to whose sounds they're most sensitive. Meanwhile, the right ears of the cellists can better hear what's going on among the violins and violas. A seemingly cultural curiosity or a historical accident – the seating plan of an orchestra – is all down to the peculiar way in which our brains have evolved.

CHAPTER 15

Einstein's Violin

THE DEEP CONNECTION between science, maths and music is reflected in the many examples of scientists who've played or composed music, and musicians who've been inspired by science or science fiction. Einstein famously said, 'Life without playing music is inconceivable to me. I live my daydreams in music. I see my life in terms of music.'

Einstein's mother, Pauline, was a capable amateur pianist and taught Albert herself. She also arranged for him to begin violin lessons when he was just five. Although he gave these up as a young teenager, impatient at having to do endless technical exercises, he continued to play and later remarked, 'I get most joy in life out of my violin.' But music for him wasn't just a hobby or a way of relaxing. It seems to have been integral to his scientific work.

While dining with Charlie Chaplin in Beverly Hills in 1931, Einstein's second wife, Elsa, recalled the events that took place one evening, sixteen years earlier. Einstein told Elsa that he'd had 'a marvellous idea'. He promptly sat at his piano and began to play, stopping now and then to jot down some notes. After half an hour, he went upstairs to his study to work on the idea in more detail. Two weeks later, his grand opus was complete, and the general theory of relativity – a new theory of gravity – was born.

Einstein had definite views about some of the great composers of the past. He loved Mozart, in particular, and became acquainted with Mozart's violin sonatas as a teen. 'Mozart's music', he said, 'is so pure and beautiful that I see it as a reflection of the inner beauty of the universe.' He admired Beethoven, too, but Mozart's music to him was special because it 'seemed to have been ever-present in the universe, waiting to be discovered'. In his study of Einstein's papers on relativity, American physicist John S. Rigden saw a parallel between Einstein's work in science and Mozart's in music. Both were 'rooted in the assumptions and conventions of their day, yet both demonstrated an intuitive leap and a new way of seeing things that lay beyond the reach of those assumptions and conventions'.[1]

Early modern scientists, such as Kepler and Galileo, didn't suppose that in achieving breakthroughs they were inventing knowledge or making connections between physical phenomena for the first time. In their view, they were simply uncovering what was already there: pre-existing patterns in the universe – glimpses, as it was assumed in those days, of God's rationality at work. Kepler, in particular, revitalised the ancient notion of the harmony of the spheres in his study of planetary orbits, linking astronomy with music in his explanation of how the planets moved. Einstein held no traditional religious views or belief in celestial soundscapes, but he often used the 'music of the spheres' as a metaphor for the quest to uncover a deep order in nature. The composers with whom he felt a strong affinity – Mozart and Bach, foremost among them – were those he believed hadn't *created* their music but rather had tuned into forms and patterns of notes that were already present in nature.

Einstein saw himself in the same vein: 'The theory of relativity occurred to me by intuition, and music is the driving force behind this intuition. My discovery was the result of musical perception.' Whenever he got stuck in his scientific theorising, he'd pick up his violin or sit at the piano and play a favourite piece, not to relax or escape from the equations and abstract concepts that were swimming around in his head but to sensitise himself to the fundamental essence of the world around him. In playing pieces by composers whom he believed had, in a sense, received their music 'ready-made' from external reality, he made himself more receptive and open to accessing deep physical truths.

Einstein was born in 1879 at a time when classical science still reigned supreme. Thanks to Galileo, Newton, Faraday, Maxwell and others, it seemed that the main task of scientists in grasping the broad principles of the universe was complete. But then cracks began to appear in the great monolith of Victorian physics: phenomena to do with hot objects, atoms and light, which simply couldn't be explained in terms of what was known at the end of the nineteenth century. The scene was set for the genius of Einstein and his ability to see to a more fundamental level than others had up to that point – a deeper vision aided by his playing of instruments almost as a cognitive tool. 'In music', he said, 'I do not look for logic. I never like work if I cannot intuitively grasp its inner unity.'

Einstein continued the long-standing tradition of finding musical harmony both a literal and a metaphorical aid in helping discern deep patterns in the way space, time, matter and energy behave. In November 1919, confirmation came of a key prediction of his general theory of relativity in the form of measurements made, during a solar eclipse, of how the Sun's gravity bends the light from background stars. On hearing the results, he treated himself in the most appropriate way: by buying a new violin.

Einstein playing the violin, 1927.

Throughout his life, Einstein owned about ten different violins, each of which he nicknamed Lina (short for 'violin'). One of them was handcrafted specially for him in 1933, the year he fled from Nazi Germany. The arrival in the United States of one of the most prominent intellectuals of the age inspired Oscar Steger, a cabinetmaker and member of the

Harrisburg Symphony Orchestra, Pennsylvania, to make a new instrument for the celebrity. Years later, Einstein gifted it to Lawrence Hibbs, the son of a handyman at Princeton University where Einstein was a resident scholar, after learning that Lawrence was a keen young player. In 2018, the Hibbs family put the violin into auction in New York where it fetched over half a million dollars.[2]

Einstein has been described as 'a professional musician's amateur musician' – an accurate, fairly talented player who practised hard, knew his parts and relished performing in ensembles for family and friends. Whether at his academic home base or on his travels, he would always have his violin with him, ready to make up a quartet for a pleasant evening of Mozart, Beethoven and Haydn in someone's living room before a small audience. His tremendous fame as a physicist also led to invitations to play with musicians who were far above him in ability – the violinist Fritz Kreiseler, the cellist Gregor Platigorky, and the Juilliard and Zoelinger string quartets.

Einstein wasn't alone as a musical scientist. A surprising number of Nobel laureates in physics, chemistry and other disciplines showed an affinity for music from an early age. Max Planck, one of the founders of quantum mechanics, was a gifted pianist and organist with perfect pitch who almost opted for music as a career over physics. Einstein spoke of Planck's 'artistic need that drove him to creative achievements'. In Berlin, the two joined forces with fellow physicist Arnold Sommerfeld and Planck's son Erwin for trio or quartet evenings in which they played for students and colleagues, such as Lise Meitner and Otto Hahn. Like Einstein, Planck turned to music to free his mind and open up new directions of thought. 'It is not logic', said Planck, 'but the creative imagination which ignites the first flash of insight in the spirit of the researcher who is advancing into dark regions [...] and without imagination new fruitful ideas do not present themselves.'

William Herschel, the German-born British astronomer who discovered Uranus in 1781, was a multi-instrumentalist and prolific composer of symphonies and concertos. Russian composer Alexander Borodin was also a research chemist who discovered the aldol reaction, important in organic synthesis; in fact, he regarded himself primarily as a scientist who composed in his spare time. It was the other way round for Edward Elgar, who made a career out of music as a distinguished composer while taking a serious amateur interest in chemistry and cryptography.

Mathematicians and musicians also share a common bond. The American popular maths writer Martin Gardner commented, 'A surprising proportion of mathematicians are accomplished musicians. Is it because music and mathematics share patterns that are beautiful?' Composer and lyricist Stephen Sondheim underscored what musicians have known for more than two millennia: 'Math and music are intimately related. Not necessarily on a conscious level, but sure.'

These connections between musicians, scientists and mathematicians extend to the present day and run deep into many forms of popular music. Olivia Newton-John's maternal grandfather was the physicist Max Born, who won the 1954 Nobel Prize for his research in quantum mechanics. Her father was the MI5 officer Brinley Newton-John, who helped break the German wartime Enigma codes. 'There's a kind of a line between music and math', she said, 'so I guess I got the music gene, thank goodness. But my mother wasn't too thrilled. She wanted me to go to university and get a degree or do something, and my father, he liked opera so he wasn't too thrilled either...'

Jimi Hendrix drew inspiration from reading science fiction while on the road. He turned the 'purplish haze' in Philip Jose Farmer's *Night of Light* into the psychedelic rock classic 'Purple Haze'. But he went beyond weaving science fiction themes into songs by composing, with his producer/manager Chas Chandler, space-rock epics that expanded the possibilities of the electric guitar and the recording studio.

Brian May did research in astrophysics before becoming sidetracked for a few decades by a project called Queen. He finally completed his PhD in 2007, aged sixty, and now divides his time between astronomy and music. Physicist and science populariser Brian Cox was the keyboard player with bands Dare and D:Ream. Guy Garvey, of Elbow, is an avid amateur astronomer. Art Garfunkel has a master's degree in mathematics. And numerous other pop musicians and rock bands, including Rush, Hawkwind and Pink Floyd, have drawn inspiration from science and science fiction. David Bowie's 'Space Oddity' was released just five days before the crew of *Apollo 11* blasted off for the Moon. In 2013, Canadian astronaut Chris Hadfield filmed himself singing 'Space Oddity' while floating in a most peculiar way around the International Space Station.

Following the publication of *Weird Maths*, co-authored with my genius student Agnijo Banerjee (winner of the 2018 International Mathematics Olympiad), I was invited to talk at the Edinburgh Science Festival. The choir I sing with in Dundee, the Noteables, thought it would be a good idea – and so it proved – to bookend the presentation with songs that we specially prepared for the occasion, including 'Calculus Rhapsody' (with apologies to Queen's better known 'Bohemian' version) and Tom Lehrer's 'That's Mathematics'. More recently, I've produced The Science Fiction Experience, a musical production that combines the sounds of symphonic rock with themes such as 'Frankenstein', 'Princess of Mars' and 'When Worlds Collide'. Another of our band's offerings is a new version of 'Mars' from Gustav Holst's *The Planets*.

Born in 1874 in Cheltenham, Gloucestershire, Holst continued a musical tradition on his father's side of the family stretching back several generations. Although he worked throughout his life as a composer, instrumentalist and teacher, his real fame rests almost entirely on a single work written between 1914 and 1917. *The Planets* had its origins in the spring of 1913, when the thirty-nine-year-old Holst was among a party of four, including the author Clifford Bax, who went on holiday together to Majorca. Bax had a taste for the esoteric – theosophy, numerology and, as it turned out during the train journey to Spain, a shared interest with Holst in astrology. Holst's fascination with the subject had already led him to read a volume called *What is a Horoscope and How is it Cast?* by the popular astrologer Alan Leo. Bax was also a passionate caster of horoscopes, and Holst was intrigued. On returning home he wrote to a friend: 'I only study things that suggest music to me. Then recently the character of each planet suggested lots to me, and I have been studying astrology fairly closely.'

Holst developed *The Planets* as, in his words, 'a series of mood pictures'. There were seven movements, each named after one of the then-known planets in the Solar System (apart from Earth) and embodying what Holst saw as the planet's character – influenced in part by Leo's astrological teachings. In 1914, Holst wrote the music for 'Mars', 'Venus' and 'Jupiter'. In the same year, Leo became one of the last astrologers to be prosecuted (though later acquitted) under the Witchcraft Act for fortune-telling. By 1917, Holst had completed the scores for 'Mercury' along with the outer planets 'Saturn', 'Uranus' and 'Neptune'.[3]

Although astrology – what he referred to as a 'pet vice' – was his starting point, Holst arranged the order of the planets to suit his own musical plan. In an early outline he listed 'Mercury' as 'no. 1', as if he intended to progress the suite from innermost planet to outermost. But the theme of 'Mars', with its distinctive 5/4 ostinato and dramatic, soaring passages, was a more obvious choice with which to open. It's sometimes supposed that the 'Mars' movement, with its clashing dissonances and driving rhythm, was inspired by the mechanistic violence of World War I, but the conflict had barely begun when Holst composed the piece. Mars, the Red Planet, has always had military connections, going back to ancient Greek and Roman times – the very word 'martial' derives from it.

The musical contrast between 'Mars' and 'Neptune', which ends the suite, could hardly be greater, though both are in 5/4 time. 'Neptune, the Mystic' is quiet, ethereal and otherwordly, its rhythm irregular and swaying, its chords occasionally dissonant and alien. The music conjures up unimaginable distances and the cold austerity of deep space. It's most chilling moments come shortly before a female chorus, screened from view, begins to sound. A dark, barely discernible rumble comes from the organ accompanied by arpeggios on the celeste. Harpists join with continuous ascending and descending glissandos before cellos and oboes add a rising melody that always just fails to resolve. At this point the voices start, intoning a soft wordless line. The orchestra falls silent and the unaccompanied voices bring the work to a pianissimo conclusion in an uncertain tonality. The chorus, as Holst says, 'fades away to nothing', as the singers slowly walk out of the auditorium into an adjoining room and a door is gradually closed behind them.[4]

Holst's daughter, Imogen, herself a composer and music teacher, remembered how the technique was perfected:

> Whenever he [Holst] was going to conduct 'Neptune' he always made his singers practise walking away while they sang their last repeated bar. On one occasion he asked me to play the cues for him at a chorus rehearsal in the Albert Hall, and I was able to see how he did it. After three repetitions of the last bar the singers slowly turned round and walked along a corridor, watching the beat of a subconductor, who led them into a room where the door was open to receive them. They walked the length of this room, still singing, while someone gradually closed the door, and at the performance they were able to go on singing until the audience's applause told them that their voices were no longer audible.[5]

Before electronic sound effects were available, Holst's way of achieving a fade-out at the end was audacious – a musical departure in more than one sense. But it wasn't quite unique. Haydn's Symphony no. 45, known as the *Farewell Symphony*, composed in 1772, used a similar strategy. Each member of the orchestra had a candle on their music stand. Towards the end of the piece, several musicians play short solos, then snuff out their candles and leave the stage. Others, without solos, gradually depart too, so that the overall sound of the orchestra steadily thins. Finally, only two violins remain to play the last notes as a barely audibly pianissimo. Haydn was sending a subtle message to his patron, Prince Esterházy – that the musicians had been away from their families too long, entertaining at the prince's favourite summer palace. Haydn himself and his concertmaster played the two remaining violins at the end of the piece, and apparently Esterházy got the hint: the day after the performance, the court and its orchestra headed for home.[6]

If Haydn had written a piece based on the planets, he would have had only six to work with (not including Earth) during his life and only five at the time the *Farewell Symphony* was written: Uranus wasn't discovered until 1781 and Neptune until 1846. When Holst composed *The Planets*, Neptune was outermost of the known worlds. Pluto took over that role in 1930, so in theory, Holst, who was still alive at the time, could have added a movement to include the newcomer in his suite. But he had no interest in doing so, having become disillusioned with how the fame of *The Planets* had overshadowed all his other work.

Fast-forward seventy years to the Bridgewater Hall, Manchester, where the Hallé Orchestra, conducted by Kent Nagano, has just finished a performance of *The Planets* with a difference. Nagano had approached the English composer Colin Matthews to write a new conclusion to Holst's work. Commissioned by the Hallé, Matthews came up with 'Pluto, the Renewer', an addendum that's now often included in performances of the suite. Matthews dedicated it to the memory of Imogen Holst, with whom he worked for many years until her death in 1984.

At first, Matthews had had mixed feelings about taking on the project, especially as 'Neptune' ends the original work so effectively. 'How could I begin again', he wrote, 'after the music has completely faded away as if into outer space?' But, in the end, the challenge proved too tempting. Starting from the imagery of solar winds blowing through the outer Solar System, Matthews created a distinctive new piece, which, nevertheless, has some echoes of the original work. Two violent, war-like passages, erupting with startling suddenness, are underpinned by a syncopated 'Mars'-like rhythm. The use of atonality gives 'Pluto' a unique, modern feel, but the 'Neptune'-like austere air of mystery is retained, as is the use of a female chorus and held note at the end after the orchestra has fallen silent. Matthews remarked, 'The movement soon took on an identity of its own, following a path which I seemed to be simply allowing to proceed as it would.' Although Einstein was no fan of twentieth-century music, one can almost imagine him smiling at that remark.

Six years after Pluto joined *The Planets* in musical form, it was controversially demoted to 'dwarf planet' status at a meeting of the International Astronomical Union. Now it finds itself in the same minor league of subplanetary flotsam as other objects such as Ceres, in the asteroid belt, and Eris, Haumea and Quaoar in the distant Kuiper belt. Still, whatever its official designation, Pluto remains an honorary planet in most people's mind, and 'Pluto, the Renewer' will no doubt retain its occasional place in the classical repertoire.

Over the past few decades, Holst's cosmic creation has served as inspiration for many film scores and rock bands. Echoes of 'Neptune', with its sense of cold remoteness, evoked by strange timbres, unfamiliar chords, and interplay of deep bass and high, sustained strings, are to be found in scenes from *Alien* and other suspenseful science fiction movies. John Williams borrowed heavily from 'Mars', both instrumentally and stylistically, in composing 'The Imperial March', first heard in *The Empire Strikes Back* and then as a recurring theme throughout the *Star Wars* movies.

With its pulsating rhythm, clashing chords and dissonant, quadruple-forte climax, 'Mars' is ready-made for transformation into heavy rock music. Pioneers of progressive rock King Crimson featured a version of it in many of their early concerts and on their live 1986 album *Epitaph*. In more modified form, employing a different staccato riff, it appears on the band's second album, *In the Wake of Poseidon*, as the track called 'The Devil's Triangle'. Emerson, Lake & Powell close out their eponymous studio album with 'Mars, the Bringer of War' and The Science Fiction Experience perform a spectacular rock version of the piece in their live shows.

The links between music and thoughts about our place in the scheme of things have never been stronger than they are today. The ancient concept of the music of the spheres is now celebrated, in updated form, at annual festivals of music and science, such as Bluedot at the Jodrell Bank Observatory in Cheshire (near where I grew up), Oxford May Music, the science tent at Glastonbury and Starmus, an international gathering that celebrates astronomy, space exploration, music and art.

A new technique called sonification is being used to turn data gathered from astronomical objects into sound and even into music. Since 2020, the sonification project at NASA's Chandra X-ray Center has been translating digital information from telescopes into musical notes, thereby opening an audio window on the cosmos. This can help in scientific work, giving a new way to analyse astrophysical events, and make data accessible to blind and partially sighted researchers in a form they can directly sense. Just as importantly, it's an extraordinary new vehicle for artistic expression and public engagement.

A composite image of the centre of our galaxy, using data from the Chandra, Hubble and Spitzer space telescopes, which was the subject of Sophie Kastner's sonification project for NASA.

Recently, the team at NASA's sonification project has been working with Montreal-based composer Sophie Kastner. The goal is to develop versions of data collected by the Chandra X-ray Observatory, Hubble Space Telescope and Spitzer Space Telescope, at wavelengths from X-ray to infrared, which can be scored and played by musicians. The pilot programme uses data from a region about 400 light-years across at the centre of our galaxy where a supermassive black hole resides. In the sonification process, computers deploy algorithms to map the digital data from Chandra, Hubble and Spitzer to sounds that humans can hear. By focusing on small sections of the image, Kastner has aimed to make the data more playable and draw the listener's attention to specific parts of the overall data set. She's called the resulting piece 'Where Parallel Lines Converge' and made the sheet music for it available online for anyone who wants to try to play it.

The full score calls for flute, clarinet and bass clarinet, piano, violin, cello and various percussion instruments, including glockenspiel, vibraphone and small tuned bells known as crotales. The piece highlights three astronomical objects or structures that appear in the chosen image: an X-ray binary, arched filaments of hot gas and the region around the central black hole, Sagittarius A*. The infrared data are mapped to a soft piano, the Hubble data to a violin and the Chandra data to a glockenspiel. As the image is scanned from one side to the other, the various regions are expressed as sounds: the gas and dust from Spitzer on mellow keys, the extended arches from the Hubble data expressed as a plucky violin and, finally, the crowded stellar space around Sagittarius A* aurally represented by a crescendo of high-energy tuned percussion.[7]

We're not used to being presented with scientific data in this way. Our brains are far more accustomed to interpreting information visually. But sonification, with its musical extension, promises to become a powerful new research tool and a source of inspiration for artists and the public at large. This is *musica universalis* in twenty-first century form.

From tunes played on bone flutes tens of thousands of years ago to evocative melodies created from the radiation of objects tens of thousands of light-years away and more – this is how far we've come in our journey as a species. Science and maths have always been at the heart of the extraordinary form of experience we know as music. But now, more than ever before, we can see clearly the threads that connect these seemingly disparate elements.

Whether we're alone in the universe or part of a great cosmic community of minds, we've come to understand that music isn't just a local peculiarity of our species. In some sense it already exists, in the same way that mathematics and the laws of physics do, outside ourselves. Music is part of the fabric of reality, an inherent component of the world around us. Together with maths and science, it forms an enduring, perfect harmony across the reaches of time and space.

Glossary

12-TET

Twelve-tone equal temperament – our modern system of dividing the octave into twelve semitones, all of proportionally equal size. Other equal temperaments include 5-TET (used in gamelan) and 24-TET (also known as quarter tones).

A440

The tone produced by an oscillation of 440 hertz. It corresponds with the A above middle C, or A4 in scientific pitch notation, and is used as a tuning standard in much of Western music, having been adopted by European classical traditions from about the mid-nineteenth century.

A CAPPELLA

Choral music with no instrumental accompaniment.

ABSOLUTE PITCH

Also known as perfect pitch, the ability of a person to identify any musical note by name after hearing it, without reference to other notes.

ACCIDENTAL

An alteration of a given pitch (i.e. making it sharp, flat or natural).

ACOUSTICS

(1) The sound characteristics of a space; (2) the science of the production, control, transmission, reception and effects of sound and the phenomenon of hearing.

AMBIENCE

The residual 'room sound' of a listening environment.

AMPLITUDE

The objectively determined volume level expressed in decibels relative to sound-pressure levels. The threshold of hearing is 0 dB, and 120 dB is the threshold of pain. All other things being equal, a wave with greater amplitude will sound louder. However, various factors affect how decibel level is related to perceived loudness. For example, we perceive very low or high sounds as being softer than middle frequencies, even if their amplitudes are the same.

ANALOGUE

Analogue representations of sound replicate its waveform while transferring it through different media. All sound is analogue, but audio can be analogue or digital.

ARTICULATION

How a note (or series of notes) is played, for example, staccato (short sounds quickly cut off), legato (smoothly connected notes) or portato (separated notes).

ATONAL

A type of harmony that intentionally avoids a tonal centre (i.e. has no apparent home key).

BAR

A common term for a musical measure.

BAROQUE ERA

The musical period from about 1600 to 1750, corresponding to a time of ornate and elaborate approaches to the arts. The Baroque era saw the rise of instrumental music, the invention of the modern violin family and the creation of the first orchestras. Composers of this era include Vivaldi, Handel and J. S. Bach.

BEATS

The interference caused when two tones of similar frequency are played simultaneously. Beats are perceived as a rapid, tremolo-like variation in volume, caused by the two waveforms alternately adding to and subtracting from each other.

CADENCE

From the Latin *cadere*, 'to fall', a succession of notes and chords that signals the end of a musical phrase.

CENT

One hundredth of a semitone, the logarithmic unit of measurement between intervals enabling intervallic relationships to remain constant while frequency values change along a different scale relative to the actual component pitches. By definition, there are 1,200 cents per octave and 100 cents per semitone, regardless of the frequency.

CHORD

A harmonic combination of two or more pitches sounded simultaneously. Chords composed of two or three notes are known as dyads and triads, respectively.

CHROMATIC

From the Greek for 'colour', a term used in one of two ways: (1) to describe notes that fall outside the seven-note scale associated with each of the twenty-four major or minor keys; or (2) to describe a twelve-note scale that contains all the notes available in the Western pitch system (i.e. a scale that uses all the 'white' and 'black' keys).

CLASSICAL ERA

The musical period from about 1750 to 1820, coincident with a politically turbulent time focused on structural unity, clarity and balance. Classical era composers include Haydn, Mozart and Beethoven.

COMMA

A small interval resulting from tuning a note in two different ways. The syntonic comma equals the difference between a pure major third and four pure fifths, minus two octaves (about 21.5 cents). The Pythagorean comma is the difference between seven octaves and twelve pure fifths (about 23.5 cents). Commas arise primarily as remainders in the construction processes of non-equally-tempered tuning systems.

CONSONANCE AND DISSONANCE

The perception of relative stability or instability, sweetness or harshness, of a chord or an interval. This depends on factors such as listening culture, musical style and historical period. For example, some chords considered consonant or stable in jazz would be considered dissonant and unstable in eighteenth-century music. Intervals generally classed as consonant include the octave (2:1), perfect fifth (3:2) and other particularly simple ratios. Intervals such as the tritone and 'wolf fifth' are commonly regarded as highly dissonant.

COUNTERPOINT

A complex polyphonic texture combining two or more independent melodies.

DECIBEL

A unit of sound intensity. It describes the relative strength of a pressure wave using a logarithmic scale (i.e. where the corresponding values for 10, 100, 1,000, 10,000, etc. are equally spaced). This roughly equates to the loudness of a given sound source.

DIATONIC SCALE

A seven-note (heptatonic) scale with five whole steps and two half steps per octave and in which the half steps are either two or three whole steps apart. Each degree of the scale has its own letter name: A, B, C, D, E, F and G. The major and the natural minor scales are examples of diatonic scales.

DIGITAL

Audio that exists in the digital realm as bits and bytes, as distinct from continuous analogue signals.

DYNAMICS

In a musical context, the subjective relative loudness or softness of a note or series of notes. It is expressed with terms such as pianissimo, mezzoforte and fortissimo. Musicians usually distinguish dynamics from volume; the former depends upon the relative range of loudness and softness in a particular musical piece or performance setting, whereas the latter is a more objective acoustic measurement.

ENHARMONIC

(1) The collapse of two pitches of a prior tuning system into one pitch with two names, for example, C sharp and D flat; (2) an enharmonic scale is one in which (using standard notation) there is no exact equivalence between a sharpened note and the flattened note it is enharmonically related to, such as in the quarter-tone scale.

ENVELOPE

In musical acoustics, how a sound changes over time as characterised by its attack, sustain and decay.

EQUAL TEMPERAMENT

The standard modern tuning system in which the octave is divided into twelve equal half steps.

ETHNOMUSICOLOGIST

A person who studies cultural aspects of musical perception, conception and appreciation, including analysis of cultural influences and assumptions regarding pitch, tone, melody, harmony, rhythm, consonance and dissonance.

FIFTH

An interval of five steps. Like the octave, the fifth creates no interference waves. In equal temperament tuning the fifths are slightly smaller than in perfect tuning – not enough to be very noticeable but enough to allow all of the half steps to be of equal size and for key changing within a piece.

FIGURED BASS

The use of numerals and symbols to specify the harmony over notated bass notes, particularly common in Baroque and early Classical music.

FORMANT

Any of several prominent frequency bands that determine the sound or phonetic quality of a vowel or consonant. Formants are the characteristic partials that identify vowels and consonants to the listener.

FREQUENCY

The number of oscillations per second, measured in hertz. We perceive differences in frequency as differences in pitch.

FUNDAMENTAL

The lowest frequency present in a periodic waveform and from which all of the overtones arise. Normally, the fundamental is the main tone we hear and the one used to classify a note's pitch. A pure sine wave comprises nothing but a fundamental.

GREGORIAN CHANT

Monophonic, non-metered melodies set to Latin sacred texts.

HARMONIC

A frequency that is an integral multiple of the fundamental. All harmonics are overtones but not all overtones are harmonics. The second harmonic is the first overtone, the third harmonic is the second overtone and so forth. A *harmonic series* is the fundamental plus the harmonics that sound at the same time.

HARMONY

The dimension of music pertaining to the succession of chords and cadences.

HEPTATONIC SCALE

A scale with seven pitches per octave.

HERTZ (HZ)

The frequency of a wave measured in cycles per second. Normal human audible range upper and lower limits are generally considered to be 20 to 20,000 Hz, based on sine-wave testing. The auditory system is more sensitive to frequencies in the range of 1,000 to 4,000 Hz.

INTERVAL

The fixed ratio of two pitch frequencies (semitones, macro/microtones); traditionally, modally defined and named (second, third, seventh, etc.) with major, minor, augmented or diminished modifiers; measured in cents.

JUST INTONATION

A category of tuning system in which all intervals are constructed using simple whole-number ratios.

KEY

The pitch centre of a particular piece or passage of music. The concept of a key is closely related to the traditional Western system of major and minor scales. Most importantly, the concept of a key implies a certain hierarchical relationship between the notes of a major or minor scale. In this hierarchy, some notes are more stable and final-sounding than others. The most stable note of the system – the tonic – is the designated centre of the key and provides the label by which the key is named.

MAJOR, MINOR

A major scale (or chord) is one that contains a major third (a note that is two whole tones above the tonic). A minor scale (or chord) is characterised by a minor third (a tone and a half above the tonic).

MAQAM

The main melodic systems of Arabic and Middle Eastern classical music, particularly in traditions arising from Egypt, Lebanon, Jordan, Palestine and Syria. Each individual maqam contains habitual melodic phrases, modulation possibilities, ornamentation techniques and aesthetic conventions, which together form a rich melodic framework and artistic tradition.

MEANTONE

A tuning system based on prioritising pure (or almost pure) thirds, by tempering the size of the fifths – narrowing them slightly from their pure ratio of 3:2. Quarter-comma meantone, which enjoyed popularity in the Renaissance and Baroque eras, preserves pure major thirds by stacking fifths of size about 696 cents (less than the pure fifth of about 702 cents).

MELISMA

A succession of many pitches sung while sustaining one syllable of text.

MELODY

A linear or 'horizontal' succession of individually sounded notes. Melody is distinct from harmony, which involves chords – a 'vertical' stacking of two or more simultaneously sounded notes.

MICROTONE

Generically, a discrete pitch interval with a value of less than 100 cents. Microtonality is a broad term that can refer to any music tuned to systems outside 12-TET (and its exact subsets). Despite the connotations of 'micro', such systems don't necessarily include intervals narrower than our familiar 100-cent semitone.

MIDI

Musical Instrument Digital Interface: a standard means of sending digitally encoded information about music between electronic devices, as between synthesisers and computers.

MODE

A scale or key used in a musical composition. Major and minor are modes, as are ancient modal scales found in Western music before about 1680.

MONOPHONY

Music based on a single melodic line, unaccompanied by any other voices or instruments.

MOTET

A sacred choral piece sung in several parts.

MUSICA UNIVERSALIS

Also known as 'music of the spheres' or 'harmony of the spheres', a philosophical concept that regards proportions in the movements of celestial bodies – in particular, the Sun, Moon and planets – as a form of music.

NOTE VALUE

The relative duration of a note. The most commonly used notes are the semibreve, minim, crotchet, quaver and semiquaver, or, in American terminology, whole, half, quarter, eighth and sixteenth note.

OCTAVE

Defined numerically, in a fixed linear fashion, as twice the frequency or half the frequency. It is based on the fundamental premise of 'octave equivalence', which relies on an assumption of a linear system (pitch perception is a non-linear process); also defined alphabetically in terms of the repetition of a pitch name while progressing linearly through the pitch sequence of a given temperament.

ORNAMENT

An extra note used to embellish the main notes of a melody. Ornaments include trills and grace notes.

OVERTONE

Any frequency that is higher than the same sound source's fundamental, comprising both its harmonic series and also any inharmonic overtones (e.g. the resonance of a snare drum). The harmonic series, also known as the overtone series, numbers these harmonic overtones by the progression of their fractions.

PARTIAL

Any of the sine waves of which a complex tone is composed, not necessarily with a whole-number multiple of the lowest harmonic.

PENTATONIC SCALE

A scale having five different notes within the space of an octave.

PERFECT INTERVAL

An interval between two notes that is consonant and stable-sounding. The perfect intervals are the unison, perfect fourth, perfect fifth and octave.

PITCH

The perceptual counterpart of frequency. As such, it is subjective and may vary from one individual, or one culture, to another. For example, the exact same up/down interval motions can sound ambiguous to different listeners, as shown by the tritone paradox and even the equivalence of the 2:1 octave.

PITCH CLASS

A note plus all of its octave duplications, including enharmonic equivalents. For example, middle C, the C two octaves above and the lowest B sharp on a piano are different pitches, but all are members of the same pitch class.

PITCH NAME

Seven repeating alphabetical (A, B, C, D, E, F and G) 'white' pitches with five flat or sharp 'black' modifiers. Each semitone is 100 cents apart.

POLYPHONIC

A musical texture in which two or more parts move independently of each other.

PSYCHOACOUSTICS

The scientific study of sound perception, typically in humans, but not in principle confined to any particular species. Psychoacoustics is an interdisciplinary field, combining concepts from neuroscience, anatomy, psychology, sociology, physics and music theory.

PYTHAGOREAN TUNING

A tuning system based on the stacking of pure fifths (3:2 ratio). This interval is used as the generator due to its consonance, offering decisive downward resolutions to every note. The method, however, presents issues with triads and major thirds, which, if created by stacking four fifths, come out at 81:64. Despite being named after Pythagoras, there is little evidence that he actually used it, although his followers did derive the string ratios of the overtone series using a monochord.

QUADRIVIUM

The more advanced division of the seven liberal arts in mediaeval universities, comprising arithmetic, geometry, astronomy and music.

RAGA

The main melodic system of Indian classical music, central to both the Hindustani and Carnatic (northern and southern) traditions. *Ragas* function like melodic mood recipes, each presenting its own ingredients, such as core phrases, note hierarchies, ascending and descending lines, and ornamentation patterns, as well as accompanying rules and guidelines for how to blend them into a coherent whole.

RELATIVE PITCH

The ability to identify or recreate a given musical note by comparing it to a reference note and identifying the interval between those two notes.

RENAISSANCE ERA

The period roughly from 1450 to 1650 in which there was a rebirth of learning and exploration. This was reflected musically in a more personal style than that seen in the Middle Ages. Renaissance composers include Josquin Desprez, Palestrina and Weelkes.

RESONANCE

The sympathetic vibration of an object or air column at a specific frequency when it is excited into motion by a sound wave of similar frequency in the immediate vicinity.

RESONANT FREQUENCY

The natural frequency at which a medium vibrates at the highest amplitude.

REVERBERATION

The sound that persists in an enclosed space, as a result of repeated reflection or scattering, after the source of the sound has stopped.

RHYTHM

The placement of musical notes and silences that occur over time.

ROMANTIC ERA

The period from roughly 1820 to 1890, an era of flamboyance, nationalism, the rise of superstar performers and concerts aimed at middle-class paying audiences. Orchestral, theatrical and solo music grew to spectacular heights of personal expression. Well-known Romantic composers include Schubert, Berlioz, Chopin, Wagner, Brahms and Tchaikovsky.

SCALE

The division of the octave into some logical or useful series of pitches. The term 'scale' indicates ascending (or descending) by predefined steps.

SCALE DEGREE

The position of a note within a scale. Scale degrees are numbered, and diatonic scales, such as the major or natural minor scale, also have named scale degrees, such as supertonic, median, subdominant and dominant.

SCIENTIFIC PITCH NOTATION

Also called international pitch notation, a method of specifying musical pitch by combining a musical note name (with accidental if needed) and a number identifying the pitch's octave.

SEMITONE

Each discrete pitch within the octave of the 12-TET tuning system.

SINE WAVE

An oscillation consisting of a single fundamental frequency and nothing else (i.e. no overtones), in other words, the simplest possible waveform – a pure sinusoidal tone derived from a simple, smooth, periodic oscillation; often produced electronically, although some 'real' sound sources (e.g. a tuning fork) come very close. Theoretically, all oscillation patterns can be modelled with combinations of sine waves.

SONIFICATION

Non-verbal sound added to complement a visual representation to make the representation easier to understand or to show new relationships in the data.

STAVE

Also known as a staff, a set of five horizontal lines and four spaces on which musical notes are written.

SWEETENED TUNING

A tuning in which some intervals are changed slightly from their 12-TET cent values in order to improve the sound in some way. On the guitar, this has a variety of manifestations (e.g. the major-third strings of open D and G are, often subconsciously, moved closer to the 'pure' third of 386 cents).

SYNTHESISER

An electronic keyboard instrument capable of generating a wide variety of sounds.

TEMPERAMENT

Tempering is the process of slightly adjusting the sizes of some or all intervals (often fifths and thirds), usually to preserve exact 2:1 octaves and accommodate other 'useful repetitions' (e.g. the capacity to modulate evenly). A tuning system is tempered if it alters the 'pure' ratios of just intonation in order to fulfil other criteria. In the 12-TET system, fifths are tempered from 702 to 700 cents, enabling a closed cycle of fifths. The practice is of particular importance for keyboards and other fixed-pitch instruments, which otherwise struggle with modulations between keys.

TEMPO

How fast a piece of music is to be played, specified by Italian words such as allegro, allegretto, lento, largo and andante, or in terms of beats per minute (BPM).

TETRACHORD

A grouping of intervals that spans a perfect fourth overall. It typically comprises four notes (*tetra* means 'four', *chordon* means 'tone'). In ancient Greek music theory, descending tetrachords of exact 4:3 ratios served as the basic unit of harmonic construction. Chains of these tetrachords, which varied in their inner notes only, formed the so-called Greater and Lesser Perfect Systems. In modern Western parlance the term still implies a fourth-like gap but can refer to any four-note adjacent sequence divided by three intervals.

THIRD

An interval of three steps; an imperfect consonance which our ears accept as pleasing.

TIMBRE

The tone colour or quality of sound, which distinguishes one instrument or voice from another. Like pitch and loudness, it is a psychoacoustical property.

TIME SIGNATURE

How many counts are in each measure and which type of note is to receive one count. The top number is commonly 2, 3, 4 or 6, and the bottom number is either 4 or 8.

TONALITY

Music centred around a 'home' key, based on a major or minor scale.

TONE

A musical sound or the intonation, pitch and modulation of a composition; also, the interval of major second (two semitones).

TONIC

Also known as the keynote, the first note of a scale or key. The tonic key is the 'home' key of a tonal composition.

TRANSPOSE

To shift a piece to a different pitch level.

TRIAD

A set of three notes, or pitch classes, which can be stacked vertically in thirds. Triads are the commonest type of chord in Western music.

TRITONE

An interval of three whole tones, dividing the octave exactly in half. Sometimes referred to as the 'devil in music', it is difficult to sing, very unstable and demands immediate resolution to a consonant interval.

TUNING

Instruments are tuned according to the laws of physics. Pythagorean tuning has all the perfect intervals (octaves, fourths and fifths) in perfect tune. Equal temperament tuning, which we use in the West today, has modified fourths and fifths but perfect octaves. This tuning is modified to allow for all the various keys composers began to use from the time of J. S. Bach onwards.

ULTRASOUND

Sound of a higher frequency than can be detected by human hearing, typically above 20 kHz.

WAVELENGTH

The distance from one wave peak to the next (i.e. the length taken up by one complete cycle of its oscillation). It is calculated as speed divided by frequency, meaning that, all other things being equal, raising the frequency leads to a shortening of wavelength.

WOLF FIFTH

A sharply dissonant interval in meantone-style temperaments, arising from the fact that a cycle of 'narrowed' fifths will not lead back to the exact octave without the inclusion of a 'stretched' fifth somewhere to compensate for the shortfall. This gives rise to a dissonant beating effect likened to a wolf's howl.

Acknowledgements

First and foremost, as always, thanks to my family – my wife Jill and our wonderful children and three grandchildren. Without their constant encouragement and support, none of this work would have been possible.

I'm immensely grateful to my editor at Oneworld, Sam Carter, and assistant editor, Hannah Haseloff, for their guidance and stimulating ideas at key moments in the development of this book.

Finally, a big thank you to my musical friends and all members of the Noteables choir, with whom I sing, for making music not only fascinating to me but also a great deal of fun.

Illustrations List

References

1. PRELUDE

1. 'The Neanderthal musical instrument from Divje Babe I cave (Slovenia): a critical review of the discussion', M. Turk, I. Turk and M. Otte, *Applied Sciences*, 10 (4), 1226, February 12, 2020.
2. 'Neanderthal bone flute music'. https://www.youtube.com/watch?v=sHy9FOblt7Y
3. 'Categorical rhythms in a singing primate', C. De Gregorio, D. Valente, et al., *Current Biology*, 31 (20), R1379–R1380, October 25, 2021.
4. 'Soprano singing in gibbons', H. Koda, T. Nishimura, et al., *American Journal of Physical Anthropology*, 149 (3), 347–355, August 24, 2012.
5. 'How music and instruments began: a brief overview of the origin and entire development of music, from its earliest stages', J. Montagu, *Frontiers in Sociology*, 2, 264256, June 19, 2017.
6. 'Monkey vocal tracts are speech-ready', W.T. Fitch, B. De Boer, N. Mathur and A.A. Ghazanfar, *Science Advances*, 2 (12), e1600723, 2016.
7. 'Evolutionary loss of complexity in human vocal anatomy as an adaptation for speech', T. Nishimura, I.T. Tokuda, et al., *Science*, 377 (6607), 760–763, August 12, 2022.
8. 'Food-associated calling in gorillas (*Gorilla g. gorilla*) in the wild', E.M. Luef, T., Breuer and S. Pika, *PloS One*, 11 (2), e0144197, February 24, 2016.
9. 'Rhythmic swaying induced by sound in chimpanzees (*Pan troglodytes*)', Y. Hattori and M. Tomonaga, *Proceedings of the National Academy of Sciences*, 117 (2), 936–942, December 23, 2019.
10. 'Sex-related communicative functions of voice spectral energy in human chorusing', P.E. Keller, J. Lee, R. König and G. Novembre, *Biology Letters*, 19 (11), 20230326, November 8, 2023.

11. 'What music makes us feel: at least 13 dimensions organize subjective experiences associated with music across different cultures', A.S. Cowen, X. Fang, D. Sauter and D. Keltner, *Proceedings of the National Academy of Sciences*, 117 (4), 1924–1934, January 6, 2020.
12. 'The form and function of chimpanzee buttress drumming', V. Eleuteri, M. Henderson, et al., *Animal Behaviour*, 192, 189–205, September 6, 2022.

2. AIRS OF AN ANCIENT AGE

1. 'The musical instruments from Ur and ancient Mesopotamian music', A.D. Kilmer, *Expedition: The Magazine of the University of Pennsylvania*, 40 (2), 12–19, July, 1998. https://www.penn.museum/sites/expedition/the-musical-instruments-from-ur-and-ancient-mesopotamian-music
2. 'The discovery of an ancient Mesopotamian theory of music', A.D. Kilmer, *Proceedings of the American Philosophical Society*, 115 (2), 131–149, 1971.
3. 'Lyre of Ur', sound example. https://www.youtube.com/watch?v=lslfCo40x7o
4. 'Kinyras: The Divine Lyre', P.M. Steele, *Hellenic Studies*, 70, 711–715, 2018.
5. 'The not so silent planets: the medieval and Renaissance concept of *musica mundana*'. https://digitalcommons.cedarville.edu/cgi/viewcontent.cgi?article=1013&context=music_and_worship_student_presentations
6. 'Greek and Roman music'. https://www.youtube.com/watch?v=18Lk_xLNdx4

3. TOWARDS HARMONY

1. 'Plainsong chant', example. https://www.youtube.com/watch?v=H4ZuHM5krp8
2. 'Ave maris stella'. https://www.youtube.com/watch?v=MNQHkipuk_s
3. 'The Music of Hildegard von Bingen', M. Hoch, *The Choral Journal*, 60 (10), 18–35, May 2020.
4. '*O Virtus Sapientiae*'. https://www.youtube.com/watch?v=lQXaDwkcQAk
5. 'Léonin, Pérotin, and the birth of polyphony at Notre Dame'. https://thelistenersclub.com/2019/04/19/leonin-perotin-and-the-birth-of-polyphony-at-notre-dame

6. 'The emergence of *ars nova*', D. Leech-Wilkinson, *The Journal of Musicology*, 13 (3), 285–317, July 1, 1995.
7. 'The quadrivium and the decline of Boethian influence', A.E. Moyer, in *A Companion to Boethius in the Middle Ages*, 479–517, Brill, 2012.
8. 'Nicole Oresme's Le Livre du ciel et du monde: A Philological View', P.F. Dembowski, 614–620 (1969).

4. A QUESTION OF SCALE

1. 'Music and perception: a study in Aristoxenus', A. Barker, *The Journal of Hellenic Studies*, 98, 9–16, 1978.
2. 'Pythagorean scale'. https://www.teoria.com/en/articles/temperaments/02-pythagoras.php
3. 'The structure of harmony in Johannes Kepler's Harmonice Mundi (1619)', in *Number to Sound: The Musical Way to the Scientific Revolution*, 173–188, M. Dickreiter, Springer Netherlands, 2000.
4. 'Problems with ancient musical scales'. https://www.cantorsparadise.com/problems-with-ancient-musical-scales-272217ca0d27
5. 'Understanding the two great temperaments: equal and meantone', J. W. Link Jr, *Journal of Research in Music Education*, 13 (3), 136–146, 1965.

5. MAKE A NOTE OF THIS

1. 'The art of musical notation'. https://www.loc.gov/collections/moldenhauer-archives/articles-and-essays/guide-to-archives/music-history
2. 'Hymn to Nikkal'. https://www.youtube.com/watch?v=tAc2KDNHEw4
3. 'Seikilos Epitaph'. https://www.classicfm.com/discover-music/seikilos-epitaph-oldest-surviving-composition
4. 'From neumes to notes: the evolution of music notation', H.R. Strayer, *Musical Offerings*, 4 (1), 1, June 6, 2013.
5. 'The legacy of Guido d'Arezzo'. https://walterbitner.com/2015/08/21/the-legacy-of-guido-darezzo
6. 'Measuring measurable music in the fifteenth century', A. Stone, *The Cambridge History of Cambridge History of Fifteenth-Century Music*, 563–586, Cambridge University Press, 2015.
7. 'Graphic notation: a brief history of visualising music'. https://davidhall.io/visualising-music-graphic-scores
8. 'Treatise, Cornelius Cardew – Kymatic ensemble'. https://www.youtube.com/watch?v=b0V9_xqaw8Q

6. RENAISSANCE AND BEYOND

1. *Fundamentals of Music*, Boethius (translated by Calvin M. Bower and Claude V. Palisca), Yale University Press, 1989. https://classicalliberalarts.com/wp-content/uploads/BOETHIUS-Bower-1989-Fundamentals_of_Music.pdf
2. 'Introductory notes on the historiography of the Greek modes', C.V. Palisca, *The Journal of Musicology*, 3 (3), 221–228, 1984.
3. 'Humanism, Italian Renaissance musical thought, and Greek tonality'. https://music.arts.uci.edu/abauer/5.2/readings/Humanism_Italian_Renaissance_Musical_Thought_Greek_Tonality.pdf
4. 'Where nature and art adjoin: investigations into the Zarlino–Galilei dispute, including an annotated translation of Vincenzo Galilei's "Discorso intorno all'opere di Messer Gioseffo Zarlino"', R.E. Goldberg, PhD thesis, Indiana University, 2011.
5. 'Galileo and music: a family affair', D. Fabris, in *The Inspiration of Astronomical Phenomena VI*, 441, 57, Astronomical Society of the Pacific, 2011.
6. 'Among others, opera's origins began with Galileo's father'. https://www.operagr.org/among-others-operas-origins-began-with-galileos-father
7. 'On the theory of the art of singing'. https://www.huygens-fokker.org/docs/stevintheory.html

7. INSTRUMENTS OF PROGRESS

1. 'First record of the sound produced by the oldest Upper Paleolithic seashell horn', C. Fritz, G. Tosello, et al., *Science Advances*, 7 (7), eabe9510, February 10, 2021.
2. 'Listen to haunting notes from an 18,000-year-old conch shell trumpet'. https://arstechnica.com/science/2021/02/listen-to-haunting-notes-from-an-18000-year-old-conch-shell-trumpet
3. 'Power efficiency in the violin'. https://news.mit.edu/2015/violin-acoustic-power-0210
4. 'Stradivari, violins, tree rings, and the Maunder Minimum: a hypothesis', L. Burckle and H.D. Grissino-Mayer, *Dendrochronologia*, 21 (1), 41–45, December, 2003.
5. 'What is a cornetto?' https://www.youtube.com/watch?v=bGcTqyC44xg

8. PITCH IN TIME

1. 'Ancient tuning methods'. https://ancientlyre.com/blog/blog/ancient-tuning-methods
2. 'Tuning the St Peter's organ, pipe by pipe'. https://www.youtube.com/watch?v=ue5TJdNDD6c
3. 'Toward a sensible tuning system for Baroque orchestras', S. Hammer, *Early Music America*, 16 (3), 68, 2010.
4. 'The science and origin of musical pitch'. https://www.togetherwithclassical.org/post/the-science-and-origin-of-music-pitch
5. 'The tuning fork: an amazing acoustics apparatus', D.A. Russell, *Acoustics Today*, 16 (2), 48–55, January 2020.

9. SOUND SCIENCE

1. 'Echeia-assisted resonance in Roman theatres', R. Godman, *Brill's Companion to the Reception of Vitruvius*, 27, 567–593, Brill, 2024.
2. 'The acoustics of ancient Greek theaters aren't what they used to be'. https://www.smithsonianmag.com/smart-news/acoustics-ancient-greek-theaters-may-no-longer-be-so-great-180965360
3. 'Acoustic diffraction effects at the Hellenistic amphitheater of Epidaurus: seat rows responsible for the marvelous acoustics', N.F. Declercq and C.S. Dekeyser, *The Journal of the Acoustical Society of America*, 121 (4), 2011–2022, April 2007.
4. 'Some aspects of the musical theory of Vincenzo Galilei and Galileo Galilei', D.P. Walker, in *Proceedings of the Royal Musical Association*, 100 (1), 33–47. Taylor & Francis Group, January 1973.
5. 'The role of music in Galileo's experiments', S. Drake, *Scientific American*, 232 (6), 98–105, June 1, 1975.
6. 'Flies, wheels and trapezoid violins'. https://www.europeana.eu/en/exhibitions/music-and-mechanics/flies-wheels-and-trapezoid-violins
7. 'Chladni plates'. https://www.youtube.com/watch?v=lRFysSAxWxI
8. 'A science superior to music: Joseph Sauveur and the estrangement between music and acoustics', A. Fix, *Physics in Perspective*, 17, 173–197, August 7, 2015.
9. 'The origins of building acoustics for theatre and music performances'. https://acoustics.org/the-origins-of-building-acoustics-for-theatre-and-music-performances-john-mourjopoulos
10. 'An acoustic history of theaters and concert halls'. https://www.acsa-arch.org/proceedings/Annual%20Meeting%20Proceedings/ACSA.AM.86/ACSA.AM.86.29.pdf

11. 'Sound now "equal" inside Royal Albert Hall'. https://www.bbc.co.uk/news/technology-47778463

10. EXPERIMENTS AND ELECTRONICA

1. 'Elisha Gray and the telephone: on the disadvantages of being an expert', D.A. Hounshell, *Technology and Culture*, 16 (2), 133–161, April 1, 1975.
2. 'The theremin: an introduction to a unique instrument'. https://www.youtube.com/watch?v=MJACNHHuGp0
3. 'Hey, what's that sound: ondes Martenot'. https://www.theguardian.com/music/2009/oct/12/ondes-martenot
4. 'Partch performs Castor & Pollux by Harry Partch'. https://www.youtube.com/watch?v=kCuYcS_Lcro
5. *Instruments of Desire: The Electric Guitar and the Shaping of Musical Experience*, S. Waksman, Harvard University Press, 2001.
6. 'The electric guitar: a documentary'. https://www.youtube.com/watch?v=PIu6yXAYT5A
7. 'The Hammond organ: genius engineering and musical icon of the twentieth century'. https://www.youtube.com/watch?v=mIjoq5dI59g
8. 'An interview with Pierre Schaeffer – pioneer of musique concrète', T. Hodgkinson, *ReR Quarterly*, 2 (1), 2, 1987.
9. 'Inside a Mellotron M400: how the Mellotron works'. https://www.youtube.com/watch?v=ByD8gH7kYxs
10. 'Max Mathews and MUSIC'. https://120years.net/music-n-max-mathews-usa-1957
11. 'Science and technology in the fine arts: music from mathematics' (review), *Science*, 139 (3549), 28–29, January 4, 1963. https://www.science.org/doi/10.1126/science.139.3549.28
12. '1969: introducing the Moog synthesiser – tomorrow's world, 1969. https://www.youtube.com/watch?v=RpuV-kpXZRI

11. DID THE BEATLES PLAY OUT OF TUNE?

1. 'James Taylor's "tuning sweetening" intonation tips'. https://gearspace.com/board/so-many-guitars-so-little-time/818338-james-taylors-quot-tuning-sweetening-quot-intonation-tips.html
2. 'Leonard Bernstein: *West Side Story* (Landmarks in Music Since 1950)' (review), L. Helgert, 531–553, March, 2011.
3. 'From Dante to Dante Sonata', in *Dante on View*, 53–64, J.E. Everson, Routledge, 2017.

12. MICROTONAL MAGIC

1. Dolores Catherino website. https://dolorescatherino.com
2. Performance of a piece by Nicola Vicentino (1555) on a reconstruction of the archicembalo. https://www.youtube.com/watch?v=JpVaAqh3nLA
3. Microtonal guitar demonstration by Tolgahan Çoğulu. https://www.youtube.com/watch?v=iRsSjh5TTqI

13. SONGS OF THE COSMOS

1. *Alien Listening: Voyager's Golden Record and Music from Earth*, D.K. Chua and A. Rehding, Princeton University Press, 2021.
2. 'Discovery alert: watch the synchronized dance of a 6-planet system'. https://exoplanets.nasa.gov/news/1771/discovery-alert-watch-the-synchronized-dance-of-a-6-planet-system
3. 'Beethoven's autopsy revisited: a pathologist sounds a final note', S.J. Oiseth, *Journal of Medical Biography*, 25 (3), 139–147, August, 2017.
4. 'Aquasonic' by Between Music. https://www.betweenmusic.dk/aquasonic
5. Between Music website. https://www.betweenmusic.dk/the-instruments
6. 'What does the universe sound like?' https://www.smithsonianmag.com/science-nature/what-does-the-universe-sound-like-180981715

14. MUSIC ON THE BRAIN

1. 'Crossmodal transfer of emotion by music', N. Logeswaran and J. Bhattacharya, *Neuroscience Letters*, 455 (2), 129–133, May 2009.
2. 'This is your brain on music: the science of a human obsession' (review), J. Robinson, *The Journal of Aesthetics and Art Criticism*, 66 (1), 91–94, 2008.
3. 'The appeal of sad music: a brief overview of current directions in research on motivations for listening to sad music', A.J. van den Tol, *The Arts in Psychotherapy*, 49, 44–49, May 2016.
4. 'Effect of music-movement synchrony on exercise oxygen consumption', C.J. Bacon, T.R. Myers and C.I. Karageorghis, *Journal of Sports Medicine and Physical Fitness*, 52 (4), 359, August 2012.
5. 'Practicing a musical instrument in childhood is associated with enhanced verbal ability and nonverbal reasoning', M. Forgeard, E. Winner, A. Norton and G. Schlaug, *PLoS One*, 3, e3566, October 29, 2008.
6. 'Longitudinal analysis of music education on executive functions in primary school children', A.C. Jaschke, H. Honing and E.J. Scherder, *Frontiers in Neuroscience*, 12, 103, February 28, 2018.

7. 'The ice-breaker effect: singing mediates fast social bonding', E. Pearce, J. Launay and R.I. Dunbar, *Royal Society Open Science*, 2 (10), October 1, 2015.
8. 'Perfect pitch: Dylan Beato'. https://www.youtube.com/watch?v=Pkx64H0F9Rk
9. 'Absolute pitch among American and Chinese conservatory students: prevalence differences, and evidence for a speech-related critical period', D. Deutsch, T. Henthorn, E. Marvin and H.-S. Xu, *Journal of the Acoustical Society of America*, 119, 719–722, February 2006.
10. 'Music can be reconstructed from human auditory cortex activity using nonlinear decoding models', L. Bellier, A. Llorens, et al., *PLoS Biology*, 21 (8), e3002176, August 15, 2023.

15. EINSTEIN'S VIOLIN

1. 'Albert Einstein: the violinist', P. White, *The Physics Teacher*, 43 (5), 286–288, May 2005.
2. 'Einstein's violin hit a high note when it sold for more than a half-million dollars at auction'. https://stringsmagazine.com/einsteins-violin-hit-a-high-note-when-it-sold-for-more-than-a-half-million-dollars-at-auction
3. 'Holst – astrology and modernism in "The Planets" ', R. Head, *Tempo*, 187, 15–24, December 1993.
4. 'How did Holst conduct "The Planets"?', A. Gibbs, *Tempo*, 66 (261), 51–58, August 8, 2012.
5. 'Introduction. *The Planets, op. 32*, by Gustav Holst', in *Collected Facsimile Edition of Manuscripts of the Published Works*, vol. 3, I. Holst, Faber Music, 1979.
6. 'Haydn Symphony no. 45 Farewell Symphony – Sinfonia Rotterdam/ Conrad van Alphen'. https://www.youtube.com/watch?v=OpD9ofCm6Ak
7. Sonification project at NASA. https://chandra.si.edu/sound/symphony.html

Further reading

Agrò, M. *Music and Astronomy: From Pythagoras to Steven Spielberg*. (New York: Springer Nature, 2023)

Ashton, A. *Harmonograph: A Visual Guide to the Mathematics of Music*. (London: Wooden Books, 2005)

Ball, P. *The Music Instinct: How Music Works and Why We Can't Do Without It*. (New York: Random House, 2010)

Benade, A.H. *Fundamentals of Musical Acoustics*. (Garden City, NY: Dover, 1990)

Bencivelli, S. *Why We Like Music: Ear, Emotion, Evolution*. (Boulogne-Billancourt:
Music Word Media Group, 2011)

Benson, D. *Music: A Mathematical Offering*. (Cambridge: Cambridge University Press, 2006)

Fauvel, J. and Flood, R. (eds.). *Music and Mathematics: From Pythagoras to Fractals*. (Oxford: Oxford University Press, 2006)

Fletcher, N.H. and Rossing, T.D. *The Physics of Musical Instruments*. (Berlin: Springer Science & Business Media, 2012)

Galpin, F.W. *The Music of the Sumerians: And Their Immediate Successors, the Babylonians and Assyrians*. (Cambridge: Cambridge University Press, 2011)

James, J. *The Music of the Spheres: Music, Science, and the Natural Order of the Universe*. (Berlin: Springer Science & Business Media, 1995)

Levitin, D.J. *This Is Your Brain on Music: The Science of a Human Obsession*. (London: Atlantic Books, 2008)

Maor, E. *Music by the Numbers: From Pythagoras to Schoenberg*. (Princeton: Princeton University Press, 2020)

May, A. *The Science of Music: How Technology Has Shaped the Evolution of an Artform*. (London: Icon Books, 2023)

Morley, I. *The Prehistory of Music: Human Evolution, Archaeology, and the Origins of Musicality*. (Oxford: Oxford University Press, 2013)

McIntyre, M.E. *Science, Music, and Mathematics: The Deepest Connections*. (Singapore: World Scientific Publishing, 2021)

Peretz, I. and Zatorre R.J. (eds.). *The Cognitive Neuroscience of Music*. (Oxford: Oxford University Press, 2003)

Suits, B.H. *Physics Behind Music: An Introduction*. (Cambridge: Cambridge University Press, 2023)

Sulzer, D. *Music, Math, and Mind: The Physics and Neuroscience of Music*. (New York: Columbia University Press, 2021)

Thaut, M. *Rhythm, Music, and the Brain: Scientific Foundations and Clinical Applications*. (Oxfordshire: Routledge, 2013)

White, H. *Physics and Music: The Science of Musical Sound*. (Garden City, NY: Dover, 2014)

Williamson, V. *You Are the Music: How Music Reveals What it Means to be Human*. (London: Icon Books, 2014)

Index

Page numbers in italics refer to photographs and diagrams. Those with the prefix 'g.' refer to terms in the glossary.

© DYLAN DRUMMOND

David Darling is a science writer, astronomer and music producer. He is the author of nearly fifty books, including the bestselling *Equations of Eternity*. Together with Agnijo Banerjee, he is the co-author of the *Weird Maths* trilogy and *The Biggest Number in the World* (also published by Oneworld). He lives in Dundee, Scotland. Learn more on his website, www.daviddarling.info, and follow him on YouTube @drdaviddarling, Facebook @drdavid.darling and Twitter @drdaviddarling.